THE GREEN EARTH CHALLENGE

Smyth & Helwys Publishing, Inc.
6316 Peake Road
Macon, Georgia 31210-3960
1-800-747-3016
©2009 by Smyth & Helwys Publishing

The paper used in this publication meets the minimum requirements of
American National Standard for Information Sciences—
Permanence of Paper for Printed Library Materials.
ANSI Z39.48–1984. (alk. paper)

Library of Congress Cataloging-in-Publication Data

Williams, Winnie Vaughan, 1932–
The green earth challenge :
the integration of faith and the
environment / by Winnie V. Williams.
p. cm.
ISBN 978-1-57312-522-2 (pbk. : alk. paper)
1. Ecology—Religious aspects—Christianity. 2. Ecotheology.
3. Human ecology—Religious aspects—Christianity. I. Title.
BT695.5.W567 2009
261.8'8—dc22
2009000067

THE GREEN EARTH CHALLENGE

INTEGRATING FAITH & THE ENVIRONMENT

WINNIE WILLIAMS

Advance Praise for
The Green Earth Challenge:
The Integration of Faith and the Environment

Winnie Williams makes "environmentalism" accessible to ordinary mortals. As a Christian, she sounds a seldom-heard moral and ethical note. She laces the text with biblical references, leaving personal interpretation to the reader.

Seldom have I seen such good stewardship of rich experiences from world travels. It is not just name-dropping of places she has been—everywhere—but careful management of lessons learned and fresh perspectives appropriated. We can all learn from Winnie's stewardship of life's blessings.

James M. Dunn
Professor of Christianity and Public Policy
Divinity School at Wake Forest University

Winnie Williams challenges us to take seriously our stewardship of God's good earth. Combining theological insights, current studies, and personal experience . . . she makes a significant contribution in calling us to live out our faith regarding our environment.

Todd Wilson
Retired minister of First Baptist Church,
Clemson, South Carolina

Dedication

To my grandsons
Carson, Taylor, Davis, and Dalton

CONTENTS

ACKNOWLEDGMENTS

I am indebted to those of my heritage who deliberately provided me sources of gratification, recreation, and pleasures through God's splendid creation. It was primarily my mother, Mamie Hardee Vaughan, who filled my spirit with delightful glimpses of nature and alerted me to God's proclamation of caring for the good earth. I am grateful for her keen insights of nature.

My heartfelt appreciation goes to my dear friend, Sylvia Titus, former professor of English at Clemson University, Clemson, South Carolina, who guided me not only through this book but through two other books as well, offering her indispensable writing skills. Rev. Dr. Todd Wilson, former pastor of the First Baptist Church, Clemson, provided valuable theological suggestions that have been most helpful. Also, my attorney son-in-law, David Vickers of Charlotte; Carolyn Crumpler of Cincinnati; and James Dunn of Winston-Salem assisted significantly in editing the manuscript.

My genuine gratitude goes to my husband, Woodie, who with near perfect patience has not only read and edited my manuscript but has also listened to endless hours of moralizing and sermonizing regarding my compassion for the environment. He has also accompanied me to some atypical but fascinating places that have provided me an enhanced global view of the ecosystem as well as the activities of humankind. Kudos to you.

I am also most appreciative to Smyth & Helwys Publishing, Inc., whose people have gently guided me through the process of publishing this book.

INTRODUCTION

I view our world as a layperson, not as a scientific environmentalist or a professional theologian. However, I have encountered God and his world's ecosystem on five continents and have developed a passion for replenishing and healing the worrisome environment that exists in most of the world's nations. Thus, from a pragmatic viewpoint, I have tried to document my observations and experiences, as well as suggest a biblical basis concerning the ethics of preserving the good earth.

Even though I have a degree from New Orleans Theological Seminary, I believe any individual can readily interpret the many biblical passages that document God's involvement in the creation of our world. However creation was accomplished, I am convinced that God was involved. I do know that God repeatedly said of the earth, "It is good—very good," and commanded us to love, protect, and use wisely the wonders of this endowment.

Many people who classify themselves as "religious" embrace the philosophy of creation care and agree that individuals, as well as the government, have a role in directing us toward a healthier and more sustainable environment. However, some of these advocates of creation care fail to participate in or defend the hunt for an ecological balance. On the other hand, some of the greatest skeptics of the green earth philosophy are the evangelical traditionalists who use God's word for justification to obliterate and consume the earth's resources.

Some of the skeptics in the evangelical camps are purists with deep religious traditions. Their tendency is to interpret the King James Version's words "have dominion and subdue it" to suggest that humans are in control and thus allowed to destroy and take advantage of every aspect of the earth for personal use. At this verse, the New Revised Standard Version[1] states, "fill the earth and subdue it" (Gen 1:28), indicating that we are to tame and keep close control over the earth, as its gardener. Some of the skeptics also adhere to the dispensationalist theory that became popular in the 1970s. This theology speculates that Jesus' return is rapidly approaching and that the earth will be abolished; thus individuals should enjoy and utilize the world's products with little forethought about the future.

However, there appears to be a slight shift of evangelicals' views, with some now accepting a more intrinsic view of creation care. Their ideas most likely will not align with secular or traditional religious environmentalist views but instead will express personal idealistic, political, and social views in their environmental agenda. In 2006, Richard Cizik, leader of the 30 million-member National Association of Evangelicals, stated in the adoption of the resolution "that every Christian has a responsibility for being good stewards of the Earth." He also advocates, "Every Christian's duty is to care for the planet and the role of government is to safeguard a sustainable environment. We affirm that God-given dominion is a sacred responsibility to steward the earth and not a license to abuse the creation of which we are a part."[2]

Another thought regarding our environmental footprints comes from my former pastor, Dr. Todd Wilson, pastor emeritus of First Baptist Church, Clemson, South Carolina, who wrote to me, "Human history shows when humans try to subdue earth for their own purposes, the image of God is defaced. When that image is obscured, dominion is impaired; when image is realized dominion is fulfilled."

I have attempted to quote Scriptures that show God's expectations for us to be responsible stewards of the earth's resources and to awaken us regarding the deep trouble that threatens our ecosystem. I have sought to reveal evidence of God's love, concern, and protection for all species of animals. I hope that the reader will derive motivation for a distinctive response to our fragile world and develop ideas for becoming more ecologically friendly.

Having traveled for twenty years to some fifty countries, my concept of America's being the wealthiest country in the world, having the world's mightiest military resources, experiencing a thriving democracy, and possessing an abundance of the world's brightest scientists and educators has often been reinforced. However, I have learned that there is much for us to glean from the masses of intelligent people who live outside our comfortable America, especially as relates to the environment. I have often heard my husband, Woodie, who also travels worldwide (but often in different directions than my journeys), say, "There are many intelligent people and ideas in the world, and they are not all in the United States." I have come to appreciate what other countries are doing to make this world a better place in which to live. Emerson said, "A foreign country is a point of comparison, wherefrom to judge his own."[3]

Globalization has affected our cultures and should enlighten everyone, including Americans, regarding the concept of a green earth. There may be more effective avenues of protecting the planet than those we presently pursue. My hope is to offer constructive, helpful concepts that we could adopt in the United States to make our world more sustainable.

For example, I was enthralled with the cleanliness of Singapore. I saw not a single piece of litter in the city. Is their clean city the result of pride, nationalism, education—or a realization that if the people drop their refuse on the ground, they will suffer punishment to fit the deed, like scrubbing the streets as well as paying a fine?

For whatever reason, many people in other countries have developed a respect for the environment. My own neighborhood is clean because each time someone drops a soda can or a fast food wrapper, a good neighbor stops to picks it up. But this act of sustaining a clean neighborhood seems to end outside the subdivision. Our community regularly collects bags of trash on the state road that passes our neighborhood. What makes the difference?

Perhaps the quandary that we face in caring for God's good earth and its habitants is how to merge our apathy with what God wishes for us to do to sustain our world. Our needs and wants seem to have walked right past God, who desires for us to integrate our faith and the environment.

Notes

[1] All Scriptures quoted are from the New Revised Standard Version (NRSV) unless otherwise noted by the Living Bible (LB).

[2] "A Blessing in Disguise," *Society Environment,* 2007, http://society.guardian.co.uk/environment/story.

[3] Ralph Waldo Emerson, "Culture," pt. 4 of *The Conduct of Life* (1860), http://classiclit.about.com/library/bl-etexts/rwemerson/bl-rwemer-conduct-4.htm.

Chapter 1

A BIBLICAL BASIS FOR HONORING THE EARTH

Black curly hair fell loosely on his forehead as the slender young man ambled through throngs of people near the ghats or steppes of India's Ganges River. He held a woven basket containing a dancing cobra, partially curled in the bottom of the basket with about a foot of its body and flat head waving in the air. I cautiously backed away from this venomous and hissing creature, trying to keep a safe distance from both the snake and its charmer. I was successful in doing so as I mingled with the meandering crowd, but it was a surreal experience.

This young man was one of the many early morning worshipers who had come to the ghats adjacent to the sacred Ganges River at Allahabad, India. Along with the snake charmer that morning were beggars, disabled people, vendors, families eating food purchased from the vendors, and drab-looking children vigorously selling chewing gum and post cards. Also there were holy men in their ocher robes, women who had come to bathe themselves and their children, and other women washing their multicolored saris and placing their clothes on the ghats to dry. Men sloshed about in the river, bathing, brushing their teeth, and washing their hair.

It was a conglomerate of people who had come primarily to drink and douse themselves in the holy waters of the Ganges. Some of the worshipers faced the morning sun with clasped hands, bowing and presenting their offerings of flowers, grain, or other foods to Ganga Ma, the Mother Ganges. Some stooped to dip their cupped hands into the river for a small amount of the holy water to drink, while others filled containers with the holy water to take to the temple to enhance their worship.

The Ganges, a sacred river, is a goddess to many of the people of India, who believe this somnolently flowing river of 1500 miles provides purity for the soul. The Hindu religion and the Indian culture embrace the philosophy that one's sins will be absolved through washing with this holy water. People travel long distances to be cleansed in this River of Heaven, which begins at the Himalayas and flows to the Bay of Bengal, passing through many cities such as Kanpur, Varansi, and Calcutta.

Unfortunately, though, in contrast to the concept that the Ganges purifies the soul, the river today is tainted with deadly waterborne diseases such as dysentery, cholera, and hepatitis and is a feeder of death. The Indian people place burnt or partially burnt dead bodies or ashes in the river, believing that the dead may reach the world of their ancestors through this process. Industries, such as leather factories,

The sacred Ganges River, Allahabad, India.

contribute to the contamination through dumping raw sewage into the water. Huge pipes dump untreated human waste into the river. At the ghats of Allahabad, I observed a dozen or so bulls in the water near the place where people were bathing and washing clothes. Since bulls are considered sacred, their bodies may find their final resting place in the Ganges as well.

Although there is some response by the Indian government, as well as some organizations, to manage the foul, contaminated waters of the Ganges River, there does not appear to be significant concern by religious groups about modifying its present conditions. Though the Ganges may be revered as a goddess, the extent of illness and death that occur due to its contaminants is unimaginable.

Before I became aware of the pollution of the Ganges River, I asked my Indian friends if they would take me for a dip in the river so that I could be "cleansed of my sins." They chuckled, knowing that I was a Christian and that I did not adhere seriously to their philosophy of cleansing sins in the Ganges River, and informed me that under no circumstances would they allow me to plunge into the contaminated water. They told me that a few of the local people acquire immunity to some of the disease-causing bacteria, but I had no such immunity. How grateful I was to my friends for their advice as I later learned about the river's contaminants.

We may think that the fouled and infected Ganges River is an example of pollution that happens only in other counties. Can such massive pollution happen in our own country? As Christians, we have a responsibility to be sensitive and compassionate about preserving the good earth God has placed in our care. Perhaps a review of biblical insights will strike a chord regarding our understanding and responsibilities toward God's gifts.

A recurring refrain in Genesis is that "all that was made is good." Whether it was the heavens, the earth, the sea, the people or animals, "God saw everything that he had made and indeed, it was very good" (Gen 1:31). But the good earth, the atmosphere, the extinction of many animals, and the welfare of millions of people is not so good in the twenty-first century.

Eastern Shores of Newfoundland, Canada.

David in the Twenty-fourth Psalm provides insight regarding the one who created this miraculous world. David states, "The earth is the LORD's and all that is in it" (v. 1). This is God's world—here, there, and everywhere—and we should be aware of his enormity, his creative power, and his unique design and acknowledge that he is the engineer and sustainer of the universe. Elihu, in conversation with Job, says, "Hear this, O Job; stop and consider the wondrous works of God" (Job 37:14).

Who better than we, incarnated beings made in God's image, to care for his magnificent creations? The psalmist states, "he founded it upon the seas and established it upon the waters" (Ps 24:2). It was all created by God and given to us to manage, but our management has not been so praiseworthy.

The sum of all that we see and feel, even that which is invisible to us, has been fashioned in God's purpose: "For in him all things in heaven and on earth were created, things visible and invisible, whether thrones or dominions or rulers or powers—all things have been created through him and for him" (Col 1:16). His creation is the sum of all things we could hope for. The universe depends on God's prevailing purpose, which provides order by him, through him, and for him.

God has not and will not abandon his creation, but his people wander far from their relationship with God and his plan for them to sustain his creation for the benefit of all. We should be committed to preserving his creation through our relationship to Christ: "He has made everything suitable for its time" (Eccl 3:11).

Paul, in writing to Timothy, talks about things that are good in themselves but bring disarray and unruliness when misused. Answering questions regarding the eating of unclean food that might offend others, Paul suggested that the people offer thanksgiving for whatever they eat and then rely on God for proper use of the food. He proposed that we submit to self-control and consideration for others in our actions: "For everything created by God is good, and nothing is to be rejected, provided it is received with thanksgiving; for it is sanctified by God's word and by prayer" (1 Tim 4:4-5). Paul suggests that we use wisdom in our actions, be moderate, and be concerned about not offending or harming others. Our acceptance of the concept that all of the earth is good suggests that we should respect and value God's creation and be cognizant of the welfare of all people.

For decades, an industry in our county dumped PCB's (polychlorinated biphenyls) into a stream that emptied into a lake behind our home. The contaminants polluted the water to the degree that today, some twenty years later, we are still advised to refrain from eating fish caught in areas of this large lake, which consists of 1,000 miles of shoreline. Our grandchildren, who love to fish, cannot understand why I do not cook their catch from the lake but instead have them throw their fish back into the water. I believe Paul would concur that those who are responsible for polluting this lake have offended our earth and have shown lack of concern for others. A recent, tentative agreement has been reached with the polluter to pay up to $22 million for cleanup, which when completed would allow the public (eventually, perhaps in several decades) to consume fish caught in the lake. Paul says, "Do not seek your own advantage but that of the other" (1 Cor 10:24).

Dr. David Rhoads, professor of New Testament at Lutheran Theological Seminary and author of "Reading the New Testament in the Environmental Age," suggests that we "review familiar Biblical passages in new ways and conceivably discover incentives for making additional constructive and effective choices regarding our ecosystem." He reminds us of the many references to nature in both the Old and New Testaments, such as the stars and the moon, the calming of storms at sea, the trees that produce fruit, and the sun as part of God's plan for creation. Dr. Rhoads also states, "There is simply one world, not humans and then the world but they are as one, not separate units. They are not even two different things kept together but simply one world."[1] He also recommends that "we look for relevant themes and patterns of thoughts shared in the New Testament writings regarding the images of new creation and the role of human steward-ship." This one world, however, has many parts and is governed by harmony through the laws of God.

Biblical teachings constantly remind us that we are the keepers, the defenders, the buffers, and the caregivers who should safeguard our planet. God creates, and we revel in God's gifts and become the hunters and the discoverers who make known his gifts. For example, the vaccine for polio was unknown until Dr. Jonas Salk discovered a powerful compound that offered protection to millions of people from the devastation of polio. New discoveries, latent in the minds of people all over the world, are revealed daily. God says that he has things for us that we do not know about, things that are yet to come. We are in a state of progression as future generations continue to dis-cover additional gifts of God's provisions.

Not only those who believe in God sense that the earth is being depleted of many of its natural resources. It has become the cry of those who observe the daily obliteration of much of the earth and the annihilation of many of the earth's species. As I travel the continents, however, I am encouraged by people of varying faiths and ethnicities who are alert to the gradual destruction of the planet. This creation belongs to God, who expects individuals to demonstrate responsibility for preserving this gift. Isaiah 66:1 says it all: "Thus says the LORD: Heaven is my throne, and the earth is my footstool."

We have lacked compassion for much of our world, which stands at the brink of desolation due to overuse of land, the pillaging of resources, and the pollution of the atmosphere. This threat is the result

"The earth is my footstool" (Isa 66:1). Nova Granada, Brazil.

of our ignorance and apathy; it is the result of individual and corporate greed, and it means we *all* must educate ourselves and commit to take action for our world.

The Wonders of God's World

God is in close proximity when I hear the pecking of a pileated woodpecker on the nearby dead pine tree or get a sweet whiff of my Abraham Lincoln rose or when the whitetail deer and her mate feast on my hostas and pansies in the backyard. God must be near when our weeping willow trembles in the wispy wind and the geese squawk as they fly confidently into the indigo sky. The feeling is unexplainable. It is similar to a hallowed caress of the spirit, like being in church, touched from inside out. God's closeness is evident in the morning when from my kitchen window I see the multi-hued sunrise kissing the trees, and the lake shines with the dithering light or when a monarch butterfly flutters its shimmering wings, waving as it passes my window. It reminds me of the depth of God's love. I know God is as near as I will let him be, and he allows me to affirm the splendor of creation.

God breathed life into the delicate orchids as they unfolded in a breathtaking garden in South Africa and in the bursting ruby-red blossoms of the 20-foot-high poinsettias. God's handiwork was evident in the plate-sized burgundy and crimson dahlias in India, and in the window boxes drooping with sun-drenched lavender and golden petunias in Amsterdam and England. God was in the vicinity when the gigantic redwood trees sprouted in California and when the Amazon's forest floor burst forth with both friendly and unfriendly plants, and when the erratic volcanoes of Guatemala erupted. Wherever I have been, sights, sounds, and events have shaped my life, and when I paused, I was able to sense God's presence. I am one of his keepers and sometimes the manager of his gifts. I have been to the edge of creation. When we get in touch with nature, we find power that restores and refreshes our souls. "Do not I fill heaven and earth?" declares the Lord (Jer 23:24).

Is it any wonder that God uses nature to illustrate his unconditional love for each of us? He clothes the fields, the grasses, and the flowers in finery and grandeur; will he do less for us? Where is our faith, he asks: "Consider the lilies, how they grow; they neither toil nor spin; yet I tell you, even Solomon in all his glory was not clothed like one of these. But if God so clothes the grass of the field, which is alive today and tomorrow is thrown into the oven, how much more will he clothe you—you of little faith!" (Luke 12:27-28). Is this not a beautiful use of nature to reassure us to be faithful followers?

Perhaps there is no psalm that affirms the magnificence of God's creative power more than Psalm 104. It is a song of glory, goodness, mercy, and wisdom, manifested through God's creation. In this song the psalmist is not only amazed by magnificent creations such as the heavens, the stars, the mountains, and the seas, but he is delighted with God's wonder in things such as flowers, birds' nests, water, grass, leaves, and raindrops. The psalmist sees the signs of providence and encourages us to accept the simple joy of ordinary life as the gift of God. So we worship, lifting our song in praise and quiet enjoyment for the small creations, as well as the magnificent ones:

You make springs gush forth in the valleys;
they flow between the hills,
giving drink to every wild animal;
the wild asses quench their thirst.
By the streams the birds of the air
have their habitation: they sing among the branches.
From your lofty abode you water the mountains;
the earth is satisfied with the fruit of your work. (Ps 104:10-13)

The psalmist ends his song with "Bless the LORD, O my soul. Praise the LORD!" (v. 35).

Many verses proclaim the beauty of God's design: "The heavens are telling the glory of God; and the firmament proclaims his handiwork" (Ps 19:1). And in Isaiah 55:10-11, "As the rain and snow come down from heaven, and do not return there until they have watered the earth making it bring forth and sprout, giving seed to the sower and bread to the eater, so shall my word be that goes out from my mouth; it shall not return to me empty." God provides the vegetation and foliage and fruit trees for our food: "Then God said, I have given you every plant yielding seed that is upon the face of all the earth, and every tree with seed in its fruit; you shall have them for food" (Gen 1:29).

A brush with God's promise.
Killarney, Ireland

Mountains Symbolizing God's Dominion

It was a Sunday morning, high in the Andes Mountains of Peru, when I found myself on an old, rather dilapidated bus accompanied by several American University colleagues. We had come to Peru for six weeks on a Fulbright Scholarship to study the history and culture of the Inca Indians and the Peruvian people. Every day, including Sundays, was packed with activities such as visiting Inca ruins, museums, churches, schools, homes, the Amazon jungle, and countless historical sites. Weeks had passed since I had worshiped in a Christian church since we seldom had the opportunity for rest or worship. I was melancholy, longing for my family, my church, and familiar places, and especially American fried chicken. As I rode on the ramshackle bus that Sunday morning, I visualized my family worshiping as we had for the past twenty-five years in the same pew, in the same church. My "holy grail" was dwindling rather rapidly, for I knew that I would not be in a church to worship for several more weeks. I was depleted of energy mentally, physically, and spiritually.

The bus, seemingly "on its last leg," followed the narrow dirt road along steep, sharp curves, slowly but surely climbing higher and higher to the peak of the mountains. I was stunned at the majestic mountains surrounding us, some snow-capped with small streams dwindling down the mountain crevices. The mostly desolate mountains were covered with rocks and large stones interspersed with small patches of vegetation and a few scrub trees. As we meandered up the desolate terrain, there was an occasional dismal shanty perched on the hillside. In the distance was a valley with an impressive lake surrounded by succulent green pastures. As we continued our climb up the steep mountains, I observed a shepherd leaning on his staff, keeping watch over his sheep, protecting them from predators as the sheep gathered bits of weeds and grass among the rocks. Immediately, in my rather despondent mood, I was reminded of the Twenty-third Psalm and recalled how the psalmist, while protecting his sheep from danger on a hillside, found fellowship with the Lord. At that moment, I, too, looked upward to my maker and experienced my own worshipful moments—just God, me, and the shepherd on the hillside. The Shepherd is ever present, even on an old bus. I was also reminded of the verse from Isaiah 6:3: "Holy, Holy, Holy, is the LORD of hosts: the whole earth is full of his glory."

Can one stand in the presence of a mountain and not feel the awesome power of God? What is it that gives most of us a sense of spiritual connection with God when we encounter mountains? Is it their magnificent height, their massiveness, their immovability, their cool atmosphere, or their simple beauty and stillness that offer our souls tranquility and peace? Whatever happens, we usually find calm and a fullness of being still and knowing that he is God.

On a summer morning while my family was vacationing in the Smoky Mountains, we dragged everyone out of bed, intending to drive to the peak of the mountains to watch the sunrise, but instead we encountered dismal patches of fog. However, what we saw was even more breathtaking than a sunrise. In a valley nearby, the clouds and fog had surrounded another mountain, and only the tip of the mountain was visible as if it were being lifted into the heavens surrounded by swirling clouds. It was a worshipful moment, and God was near. We all reveled in the beauty and complexity of yet another unspeakable encounter as the morning sun polished the clouds.

Mountains have played significant roles in many of the great experiences of the Bible, both in the Old Testament and the New Testament. Remember Moses being given the Ten Commandments at

A testament to nature. Glacier National Park.

Mt. Sinai, Elijah on Mount Carmel, Jesus making a triumphal entrance from the Mount of Olives, and the transfiguration taking place high on a mountain. Isaiah gives reassurances of the promises of God: "For you shall go out in joy, and be led back in peace; the mountains and hills before you shall burst into song, and all the trees of the field shall clap their hands" (Isa 55:12).

God often uses the mountains as a symbol of his immense power, especially in Psalms. The psalmist asked, "Who shall ascend the hill of the LORD? And who shall stand in this holy place?" (Ps 24:4). Another psalm acknowledges God as the creator of the universe: "I will lift up mine eyes to the hills—from where will my help come? My help comes from the LORD, who made heaven and earth" (Ps 121:1-2). The verses indicate a relationship with God who not only provides us with help but is also the one who formed the mountains and the fullness thereof.

I love mountains. I've climbed a few, but I mostly enjoy them from a distance. However, I remember riding in a small boat for four hours one morning, enduring frigid, blustery weather to reach a mountain where a settlement of Quecha Indians lived. This mountain was located in the waters of Lake Titicaca, which borders Brazil and

A spiritual endowment, Southern Ireland.

Peru and rises to a height of 13,500 feet. The lake, which has the highest elevation of any lake in the world, surrounds the large mountain and gives the appearance of the mountain bursting out of the lake. Because of the extreme height of the mountain, our guide carried a small oxygen tank in case of emergency. I followed the guide closely as we climbed the mountainous island because I wanted to be near him if I needed a little extra assistance.

We lunched on sandwiches at the foot of the mountain before we began our hike. A candy bar was part of my lunch, but I decided to squirrel it away for later in the day, as no other food would be available until we returned to the boat. As we climbed the rocky path up the mountain, a small Peruvian Indian boy darted from bushes in front of me. He placed his open hands together and lifted them up to me, a clear signal of his hunger. I did not wish to part with my candy bar. It was my only source of energy, and I resented momentarily the predicament of feeling guilty for not giving the candy bar to the child. My better instinct prevailed, though, and I lifted the candy from my jacket pocket and handed it to the hungry boy. Without a word, he took the candy and darted into the bushes as quickly as he had appeared. I knew I had done the right thing.

Climbing this mountain and discovering the Quecha tribe was an awesome spiritual adventure. They survive by the toil of their hands on this mountainous island without the intrusion of modern culture. Nature provides them with their needs; they even weave their clothes from alpaca wool.

Other pristine mountains that beheld the glory of God's creation were those in Glacier National Park. Recently my husband and I spent several days observing the marvelous snow-capped mountains and were reminded that as the mountains burst from the earth, they became a testimony to God's creation. These mountains are named for the rivers of ice that cover their peaks, carving spectacular alpine landscapes. As the snow melts, it forms steep waterfalls that drop into the crevices and flow into the opaque emerald lakes. It is a diverse landscape, and every spot is an awesome view of mountainsides, waterfalls, trees, and native flora surrounding the valleys filled with pure, unspoiled river water. It is God's creation at its best, where creatures are on the prowl and wild birds cloud the unblemished blue sky. It is a place where one can dwell indefinitely, just to be in the presence of the Lord.

Other mountains that have lifted me to great spiritual heights are the Himalayas, the Swiss Alps, the glacier-covered mountains of Alaska, and the smaller mountains of Haiti. Whether the mountains are huge or small, sandy or rocky, green or barren, the psalmist says we should "Praise the LORD from the earth . . . mountains and all hills, fruit trees and all cedars" (Ps 148:7, 9).

The Mysteries of Nature

In my rush to get things accomplished, I have missed much of the beauty that lies within my reach. My ears often failed to hear the chatter of chipmunks or the flutter of a hummingbird's wings because I was too busy pondering my next challenge. I have missed the unfolding of blooms and branches waving in the wind or failed to see the blades of grass twisting out of the earth. I have missed touching the velvet moss-covered stone or seeing the reflection of shimmering sun rays glinting off a seashell—all overflowing with God's beauty.

Our family's haven is a small cabin located on several acres of forest at the foot of the mountains in North Georgia. It is a throwback to my childhood when my family lived in the low country of Mississippi. Our neighbors include whitetail deer that munch on my scattered hostas and mongo grass, squirrels that thrust the shells of their nuts over the cabin deck, a menagerie of birds that fight with the squirrels at the bird feeder, cottontail rabbits, wild turkeys, and an occasional black bear scavenging for food. The trees dance as the winds bend their limbs, and the only sounds are the melodious songs of the birds and the splashing echoes of water as it tumbles over rocks in a nearby stream. It is a spot so quiet that we can hear nature speaking. Sig Olson said that "without stillness there can be no knowing."[2] The cabin retreat has become our place of knowing, and we take the singing of nature with us as we leave. It is our hope that logging or developments or drillings do not chase the woodlands and our haven into exile.

We Americans like to sense that we have a monopoly on most things, especially that which is picturesque, pleasing, and valuable. We do have a hefty share of the natural world's resources, but these resources are not limited to America. Other countries share in God's beautiful creations, and some of these countries surpass America in honoring and preserving their gifts.

Some of my most breathtaking and awesome moments of observing God's good earth have been in developing countries. For example, India, encumbered with massive poverty, has some of the finest natural landscapes I have seen. Portions of the country have long growing seasons that enrich their gardens, producing a continuous unfurling of giant showy blooms of flowers and vegetables. Their flat, low-lying lands, often tilled by hand, produce magnificent landscapes, and many of their natural phenomena such as mountains, rivers, and caves are captivating. For example, I waded through a cave in northern India for perhaps one-fourth of a mile, sloshing in one to two feet of frigid water in near darkness to experience one of nature's most magnificent formations. The adventure was challenging but well worth the discomfort to discover the river-like stream meandering through the cave to an island where a Hindu priest stood in his ceremonial robe, accepting sacrificial offerings from Indian worshipers. Experiencing this rare cave left me in awe of nature's phenomena.

It is not unusual for one to travel many miles to view a magnificent garden, waterfall, or other scene of nature. My search for beauty and solitude in nature led me to the majestic gardens of England this past year, where I observed some of the grandest and most elaborate gardens that exist. Among those that appealed to my soul were the gardens of Buckingham Palace, the Chartwell Gardens of Winston Churchill's home, and the homey gardens that surround the thatched-roofed home of Anne Hathaway. God's manifestation in these gardens is almost too wonderful for words.

Some of the places of grandeur in our world are being threatened through exploration and environmental factors such as global warming. For example, we observed the warming effect in the Norwegian glaciers, which are melting at an alarming pace. Other areas with melting glaciers are Alaska, Iceland, Greenland, the Arctic, and Antarctica.

Formerly ice-covered mountains are now brown due to the melting of the glaciers. In addition to damaging wildlife habitats that rely on the ice, melting glaciers cause the sea levels to rise. If we fail to correct the warming trend, coastal and low-lying areas of the world will be engulfed in water and most likely will never be restored to their original beauty.

Americans contribute significantly to the current state of the earth. Jimmy Carter wrote in *Our Endangered Values* that "America is by far the world's leading polluter, and our government's abandonment of its responsibilities is just another tragic step in a series of actions that have departed from the historic bipartisan protection of the global environment. Our proper stewardship of God's world is a personal and political moral commitment."[3]

What kind of impact will global changes make in our lives? For each person it will be different, but for me it has taken almost a lifetime to look beyond my own space to the space that God constantly unfolds before my eyes. He has led me to every corner of the world in discovering his creations and the miraculous replenishment process of his plans. I now seek to be more appreciative and protective of God's creation, whether gardens, glaciers, lakes, mountains, air, or seas. I want to grasp that beauty, internalize it, and offer praises for what I discover. Beauty is everywhere. We need only our eyes and minds to perceive what has been made and given to us for our use. The Psalmist recalls the work of the LORD, who said, "For the world and all that is in it is mine" (Ps 50:12b). Hopefully we will not squander this gift.

Questions for Reflection

1. God stated that when the earth was created, it "was good." Is it "good" today? If so, in what ways? If not, why not?
2. Are we on a path to discover the gifts God has for us?
3. What in nature gives you a sense of spiritual connection to God?
4. Do you feel that God is pleased with our management of the earth? In what ways has our management benefited society? How has it harmed society or the world's disenfranchised citizens?

5. What environmental concerns confront your community? What is being done about these problems?

Notes

[1] David Rhoads, "Reading the New Testament in an Environmental Age," *Currents in Theology and Mission* 24/3 (1997) 259–266. Available online as "Reading the New Testament in the Environmental Age," http://www.webofcreation.org/Articles/rhoads.html.

[2] Jim Brandenburg, "Next stop in the American Landscape series: the beguiling wilds of northern Minnesota," http://ngm.nationalgeographic.com/ngm/0306/feature2/index.html.

[3] Jimmy Carter, *Our Endangered Values* (New York: Simon & Schuster, 2005), 177.

Chapter 2

THIS LAND OF OURS: CLIMATE CHANGE

There seems to be a general consensus among most Americans that rapid climate change is apparent, even though some skeptics advocate that it is a hoax. At last, though, most people are realizing that our world is experiencing a major warming trend. To what degree climate change or global warming is due to natural climatic conditions versus manmade warming has not been validated specifically. However, climatic scientists agree that much of global warming results from humans releasing pollutants into the air and water.

Apprehension regarding environmental issues is not a new phenomenon, even though it has captured the recent attention of the media, environmentalists, laypeople, and politicians to a greater degree than previously noted. History records that in 1306 King Edward I of England banned the burning of sea coal in London because it produced so much pollution. During the following centuries, Great Britain continued to seek to reduce excessive smoke. They were not alone in their concern for air pollutants. During the American Industrial Revolution, factories emitted enormous amounts of pollution, and cities such as Chicago and Cincinnati enacted clean air legislation as early as 1881.[1]

However, air pollution didn't cause alarm in America until the 1940s, when cities such as Los Angeles became inundated with smog. It was another decade until the federal government passed the Air Pollution Control Act of 1955, the first of a series of clean air acts that continue today but with numerous modifications. The purpose of this act was to identify air contaminants and to provide research and means whereby air pollutants could be reduced. Most of all, this act alerted the nation to the hazards related to air pollution.

There were additional amendments to the Air Pollution Control Act (which became the Clean Air Act of 1963) such as setting emissions standards for power plants and steel mills. Further amendments were passed in 1965, 1966, 1967, 1969, and 1977. These acts regulated auto emissions, expanded local air pollution control programs, and established air quality and compliance deadlines. By 1970, more demanding standards were set. However, the auto industry, coal industries, and other polluting entities advocated that the enforcement of these limitations would impose economic hardships on them. Thus, Congress reneged and extended deadlines. During the 1980s there were no revisions to the Clean Air Act, but in 1990, due to growing environmental concerns, the federal government passed the Clean Air Act of 1990. Issues addressed were air quality standards, motor vehicle emissions, alternative fuels, toxic air pollutants, acid rain, and stratospheric ozone depletion. Primarily, this law was to strengthen and improve existing regulations.[2]

In spite of numerous positive provisions of the Clean Air Act, it came under attack by environmentalists, who claimed that it was laden with loopholes that actually allowed industries to avoid restricting air pollutants. Controlling emissions became a political football, and industries were able to dodge critical environmental regulations that would have enhanced air quality.

The United Nations has adopted two worldwide treaties to address global warming: the Rio and Kyoto (2005) agreements. The Rio conference took place in Rio de Janeiro, Brazil, in June of 1992 and was a plan of action dealing with the environment and sustainable management of forests. The agreement was adopted by more than 178 governments. The commitments to these environmental principles were reaffirmed at the World Summit in Johannesburg, South Africa, in September 2002.

The Kyoto agreement of 2005 was designed to reduce emissions of carbon dioxide and five other greenhouse gases and to engage in emission trading. Emission trading allows a company that exceeds a gas cap, set by a governmental agency, to purchase credits from a company who emits less than their designated caps. It is a system that rewards those who decrease emissions and in essence places a fine on those who emit excessive gases. One hundred and twenty-five countries signed the agreement, but the United States and one other country refused to do so. The major reason given by the United States for not signing was that it would be "too costly" for industries of the United States, who claimed that such an agreement would significantly lower profits for their stockholders and affect the nation's economy. The United States government concurred.

When a country prefers to shore up individual profits rather than unite with others for the health and productivity of the world (of which we are part!), we have a considerable moral issue to confront. Luke has a word regarding greed as revealed in the story of the rich young ruler who desired to fill his barns to overflowing. God said, "Take care! Be on your guard against all kinds of greed; for one's life does not consist in the abundance of possessions" (Luke 12:15).

Perhaps the most rousing environmental evidence ever produced regarding global warming was the statement released by the Intergovernmental Panel on Climate Change in February 2007. The panel, composed of scientists (and including some skeptics) and representatives of 113 governments, presented evidence of global warming and unanimously stated that conditions as they presently exist are worsening. "The report," according to co-author Andrew Weaver of the University of Victoria, "was based on science that is rock-solid, peer-reviewed, conservative and consensus as the result of years of research. There was little debate on the strength of the wording regarding human activity as most likely to blame."[3] This was a powerful statement on the environmental issue by the world's leading scientists, who noted that there are already visible signs of global warming, such as rising water, killer heat waves, worsening droughts, and stronger hurricanes.

Defining Global Warming

So what is this climatic change or global warming that appears to be ruinous and destructive to God's earth? It exists when a chemical reaction occurs as the sunlight impacts pollutants, damaging the ozone layer that lies in the atmosphere ten to thirty miles above ground. Ozone is a form of oxygen that occurs naturally and provides a shield from the sun's ultraviolet radiation. Global warming is the result of gases produced from burning gasoline, diesel fuel, natural gas, and coal: carbon dioxide (mainly) and methane, nitrous oxide, and water vapors (to lesser degrees). When released into the atmosphere, these gases block thermal radiation (i.e., heat) from leaving the earth—hence the term "greenhouse gases."

At ground level, the ozone levels or smog are highest at midday, especially if the air is stagnant and the sun is intense, baking the chemicals in the air. People with chronic pulmonary disease such as asthma, bronchitis, and emphysema should avoid exercising in midday when the ozone level is at its peak. If the air is polluted, there may be twenty times more pollutants entering the body when exercising at the hottest part of the day. Carbon monoxide (usually from auto exhaust) reduces the blood's ability to transport and release oxygen in the body. The Department of Health and Environmental Control provides a scale to determine current ozone levels: green (0–50) meaning good; orange (101–150) meaning unhealthy for vulnerable people; and purple (201–300) meaning very unhealthy.[4] The scale is useful for anyone who wishes to plan outdoor activities.

Effects of Global Warming

We as Americans make up only 2 percent of the world's population, yet we consume 25 percent of the world's fuel as a result of our luxurious lifestyles. Compared with most people of the world, we own big cars, live in huge heated and cooled houses, consume whopping amounts of food, use utilities excessively, and possess colossal wardrobes. All of these activities requiring fuel and other natural resources result in Americans contributing more than our share to global warming. What a high price the world pays for our sumptuous lifestyles!

Our lavish lifestyles remind us that we may be similar to the rich young ruler. We live in a culture where we have an insatiable appetite for material things. Producing these possessions requires resources, energy, and transportation. Our moral challenge is to consider our wants versus our needs, to figure out why we equate our happiness— even our "American-ness"—with our ability to consume, and then to realize that we simply don't need so much "stuff." We must re-envision American lifestyles so that we may participate in the reduction of greenhouse gases before conditions become irreversible. We must want to make a difference before changes will occur.

Global warming is a threat to humans, animals, and all of God's creation. The National Academy of Sciences states that the last decade of the twentieth century was warmer than any comparable period in the past 400 years and that our planet and ecosystem is at risk. Earth has not experienced such rapid temperature changes in a thousand years, according to the National Academy.[5] Not only does global warming deplete the earth of its natural resources, but it also creates health concerns such as skin cancer and respiratory problems, as well as environmental concerns such as the melting of glaciers and flooding, and powerful hurricanes like Katrina. Global warming is playing havoc with animal migrations and changing the habitat of fish, and causing some varieties of vegetation to overwhelm native flora.

Already global warming is damaging the ecosystem of undeveloped countries that try to cope daily with barely enough food for their people to survive. For example, the droughts experienced in Africa, the loss of wetlands in areas such as India, and melting glaciers that cause the rising of sea levels, thereby submerging land, has been especially costly for the poor, who are directly dependent on the earth for their livelihood.

Demands of population growth contributes to global warming.
District of Uttar Pradesh, India

We have a scriptural obligation to be concerned for the poor: "You shall love your neighbor as yourself. But if you show partiality, you commit sin and are convicted by the law as transgressors" (Jas 2:8-9).

The Sierra Club[6] reported that changes in global warming could produce devastating, irreversible effects. The question is whether the global warming trend is moving at such a rapid pace that we may even now be helpless to reverse or even slow the trend.

During my travels I have observed extreme air pollution in cities like Shanghai, Moscow, Bangkok, and numerous cities in India. With the stifling exhaust fumes in these cities, it is necessary to cover the nose and mouth with a face mask to filter some of the dust and pollutants. One's eyes often smart from the pollutants. Nitrogen dioxide and sulfur dioxide have been shown to irritate the eyes, as well as constrict the body's air passages. Usually the pollutants are the result of emissions from industry and auto exhausts reacting to sunlight and heat. These pollutants are sometimes referred to as smog, which we can find in abundance in cities like New York and Los Angeles.

While I was in Bangkok, my eyes burned and my nose felt itchy and irritated as we struggled through congested traffic, often finding ourselves in traffic jams for long periods. This city of 10 million people appears to have had inadequate traffic control until recent years, with no subway system or carpool lanes. Because of the crowded roads, one could not expect to arrive at any destination on time. My missionary friend in Bangkok walked her little girl some eight to ten blocks to school each morning until the little girl developed a respiratory disorder. Her doctor suggested the child no longer be allowed to walk to school along the city streets because of the pollution. This child was affected by air pollution as a result of simply walking back and forth to school, but imagine the fate of people such as policemen and taxi and bus drivers working eight-hour days in such polluted areas. This same scenario can and does happen in major cities in the United States.

As my husband and I traveled to Alaska and Glacier National Park, we observed firsthand some of the places of grandeur in our country that are being threatened by global warming. In both places we saw evidence of the warming trend as the tops of former ice-covered mountains lay bare and brown. At Glacier National Park, we saw photos of a mountain range twenty years ago covered with snow and ice and photos of the same mountains today—without ice and snow at the same time of year. Inch by inch, the glaciers are melting, presenting a formidable devastation for the ecosystem, including the flooding of streams, the rising of the sea, and the damaging of wildlife habitats.

Scientists at the park predict that if there is not a correction in the warming trend, pieces of America will disappear or never be restored to their original natural beauty. In 1850, there were 150 glaciers in Glacier National Park; today there are approximately 35. Without a major reversal of global warming, scientists predict that by 2030, not even one glacier may be left. From Norway to Alaska, 90 percent of the world's glaciers are shrinking. Last year we were in Newfoundland and Labrador, Canada, and talked with a lighthouse attendant regarding icebergs that travel along the coastline each year. She stated that normally there are 500 icebergs floating in their vicinity, but this past year they saw only 3, as most of the icebergs melted prior to reaching Labrador's coast.

Winnie explores glaciers in Alaska.

While in Alaska I witnessed my first "calving" of an iceberg as large chunks of ice collapsed into the ocean. It was awesome and frightening as a portion of a glacier, perhaps the size of a house, broke off and collapsed into the water with a thunderous splash. Many areas over the world are experiencing frequent "calvings" due to rising temperatures. The highest rise in temperature has been in Antarctica, where it has been reported that temperatures have risen 4.5 degrees over the past 50 years. In February 2008, the Wilkins Ice Shelf, "a broad plate of permanent floating ice [of more than 5200 square miles] on the southwest Antarctic Peninsula," began to disintegrate because of the warming temperatures.[7] The 2002 break-up of the millennia-old Larsen-B ice shelf gained worldwide attention when it disintegrated in less than thirty days off the West Antarctic Peninsula.[8]

According to NASA, the polar ice cap has shrunk by 250 million acres—an area the size of California, Maryland, and Texas combined—and is receding at the rate of 9 percent each year. At this pace the entire polar ice cap could disappear altogether by the end of the century.[9] As indicated in a report released by the Arctic Council, 300 scientists from 8 countries, including the United States, agreed that manmade global warming is already having a disastrous impact on people and wildlife in the Arctic.[10] Polar regions are bellwethers—like canaries in the mine—a foreshadowing of what may come for every living thing on earth.

Frank Crowder, conservation chair of the South Carolina chapter of the Sierra Club, states, "Increased conservation, energy efficiency, and use of renewable resource will not happen through new laws and policy alone. While those are necessary, there also must be a fundamental shift in awareness, attitudes, and values that connect directly to experience, learning, and thoughtful consideration of a better way to live on our planet."[11] It is Crowder's hope that we can build a humane global society, honor human rights and dignity, and restore and protect the natural environment.

The Moral Challenge

The environment of the earth is affected by the quality of air, water, plants, soil, animals, and people within its boundaries. When the Creator made the world and "saw that it was good," everything was in harmony. The ecosystem was perfect in every sense of the word, but

Glacier National Park meltdown.

because of the imperfections of human beings, much of the world's "good earth" is disappearing. When individuals seek to steer progress toward their own interests, the environment is thrown into disarray. The earth groans because we have failed to nurture what has been provided for our use.

Humans, created in the image of God, are set apart from animals and other parts of the earth. Although God placed us in command of the earth, we are not given freedom to abuse those gifts. Instead, in God's word, we are given responsibility to demonstrate stewardship for land and animals. "The Lord placed the man in the Garden of Eden as its gardener, to tend and care for it" (Gen 2:15, LB).

In Leviticus 25:4, God addressed Moses on Mount Sinai about the matter of being a steward of the land. He instructed the Israelites to grow and gather crops for six years, but on the seventh year, "There shall be a sabbath of complete rest for the land." He further instructed that cattle and wild animals should be allowed to graze on the land during the Sabbath. To treat our world properly, we need to have an understanding of the elements and their effects not only on people but also on animals, the land, and the oceans. Unless we become informed, we will not act in a manner appropriate to preserving our earth.

In a Christian denominational newsletter, "Christians and the Environment," the writer states his biblical concept of an ecological crisis:

> The natural resources on which man depends are beginning to be depleted or despoiled by the activities of man. This is illustrated by the history of God's chosen people in Israel. They were given a good land to enjoy—"a land flowing with milk and honey." (Deut. 11:9) They were also given a set of guidelines for its use (Lev 25) and were warned what would happen if they did not obey these rules. Despite the warnings they did not obey, and by Isaiah's time, the land had become derelict and lay in waste. (Isaiah 33:9) The land had become barren, the tree cover had disappeared, and the soil was eroded away. Had the people followed God's commandments, they then would have had ample fruit and crops.[12]

There is a spiritual and moral imperative that everyone, especially those who cherish God's word, should labor together in providing the best policies for our environment through the reduction of pollutants. We must be willing to gather information, then analyze and internalize the findings in order to address environmental issues. Individually and collectively, we must be forceful in making changes. It is necessary for us to pick up the challenge and become advocates to save our planet—if not for ourselves, then for our children and grandchildren. Together we can and must be intentionally involved. The consequences of not caring or acting on the commandments of the Bible are perilous. Can we heal the wounds of the earth within our own lives?

A church in the wildwoods. Labrador, Canada.

Questions for Reflection

1. To what extent has the evidence on global warming been politicized? Given the evidence, how do our actions contribute to these warming conditions?
2. There are individuals and groups, including religious entities, who oppose efforts to curb global warming. What is their rationale?
3. What are the consequences of global warming for animals, such as the polar bear, and for the fish of lakes and oceans?

4. What are the political barriers to enacting laws that protect the environment? What steps can individuals take to participate in this political process?

5. What effects would America's signing the Kyoto agreement have on this country? On other countries?

6. Is emission trading an effective strategy to combat global warming? Why or why not?

7. Has your understanding of caring for God's earth changed over the years? If so, how? If not, why not? What is the relationship between your faith and your view of the environment?

Notes

[1] James Fleming and Bethany R. Knorr, "History of the Clean Air Act," Colby College, Waterville ME, http://www.ametsoc.org/Sloan/cleanair/index.html, January 2007.

[2] "A Look at United States Air Pollution Laws and Their Amendments," Environmental Protection Agency, ametsoc.org/sloan/cleanair/cleanairlegisl.html, 2 February 2007.

[3] "Report strongly links humans to warming," Greenville (SC) News, 2 August 2007.

[4] Mike Foley, "The Air," Greenville (SC) News, 7 February 2006, p. 1.

[5] South Carolina Wildlife, 8 July 2006, 50.

[6] "The Basics of Global Warming," Sierra Club, http://www.sierraclub.org/energy/overview/ (30 December 2008).

[7] "Antarctic Ice Shelf Disintegration Underscores a Warming World," National Snow and Ice Data Center, http://www.nsidc.org/news/press/20080325_Wilkins.html (30 December 2008).

[8] Ibid.

[9] "The Heat Is On: A White Paper on Climate Action," Environmental Defense (2004), http://www.edf.org/documents/3777_TheHeatIsOn.pdf (30 December 2008).

[10] "Report strongly links humans to warming."

[11] Congaree Chronicle 29/6, Sierra Club SC, chs. 11–12, 2006, 1.

[12] Tony Campolo and Gordon Aeschilman, "Christians and the Environment: The Biblical Basis," www.spiritrestoration.org/Church/All%20About%20Church%20Articles/Christians%20and%20the%20Environment.htm (2 December 2008).

Chapter 3

THIS LAND OF OURS:
THE FLORA

Rachel Carson's *Silent Spring*, first published in 1962, set into motion the initial series of events that heightened awareness regarding environmental pollution and its effects on land, air, and water.[1] At risk are not only animals, insects, and sea life but humans as well, Carson warned. As a scientific writer and biologist, she predicted that destruction lay ahead if we did not alter our approach in caring for the earth and its inhabitants. With hindsight, we see just how noteworthy her predictions were.

Carson's research analyzed the outcomes of the extensive use of DDT and its implications for the environment. Upon publication, *Silent Spring* was the most enlightening book written regarding the ravaging of the ecosystem. Carson presented a view of the damage chemicals caused and revealed how the soil becomes contaminated, affecting the worms living in the soil, which in turn result in a deadly food source for birds. Eventually, the poison passes through the food chain and reaches the water, where it kills both fish and water vegetation. Carson was on the right track even forty-five years ago, but somehow in these past years, her predictions have not resonated with the public. There has been a rapid development of new and more potent chemicals contaminating the earth and, eventually, humans. As

Albert Schweitzer aptly stated, "Man can hardly recognize the devils of his own creation."

God's Green Bible

Schweitzer's suggestion that we lose sight of the harm we create is apt. Why does this happen? We seek to improve our crops, control pests, and the like, but in the wake of our solutions are the devils that cause more far-reaching harm. Perhaps we are victims of our own greed, ignorance, apathy, and failure to love our neighbor as ourselves. A larger question might be this: when we contaminate, pollute, defile, poison, and disrupt the ecosystem, *are we sinning?*

Robert Parham, executive director of the Baptist Center for Ethics, writes in *Reading the Green Bible* that the "Bible is God's green book, staking out the divine imperative for the earth's care. Even the red-lettered Bible, the one that has what Jesus said in red letters, is really a green Bible. In fact, Jesus' great Commandment (Matthew 22:37) has green all over it." Parham refers to "You shall love your neighbor as yourself" as the "critical verse that requires us to care for people throughout the world now and across time to whom we have a moral obligation."[2]

Old Testament writings give specific direction regarding the caring for that which God has placed in his creation. For example, soldiers are commanded, "If you besiege a town for a long time, making war against it in order to take it, you must not destroy its trees by wielding an ax against them. Although you may take food from them, you must not cut them down" (Deut 20:19). Also, the Lord instructed Moses to inform the people of Israel that when they came into the land of Canaan, which he was giving them for their homeland, they "shall not defile the land in which you live, in which I also dwell; for I the LORD dwell among the Israelites" (Num 35:34). The Lord God speaks with tenderness and caring for the earth when he says,

> I myself will take a sprig from the loft top of a cedar; I will set it out. I will break off a tender one from the topmost of its young twigs; I myself will plant it on a high and lofty mountain. On the mountain

height of Israel I will plant it, in order that it may produce boughs and bear fruit, and become a noble cedar. Under it every kind of bird will live; in the shade of its branches will nest winged creatures of every kind. All the trees of the field shall know that I am the LORD. I bring low the high tree, I make high the low tree; I dry up the green tree and make the dry tree flourish. I the LORD have spoken I will accomplish it. (Ezek 17:22-24)

Another Old Testament example of nurturing creation is shown when Hosea tells the Israelites that because of their unfaithfulness, lack of love, and acknowledgment of God, "The land mourns and all who live in it languish; along with the wild animals and the birds of the air, even the fish of the sea are perishing" (Hos 4:3). How prophetic for our time!

In addition to these verses, we find Jesus using nature as an illustration to help the disciples understand the kingdom of God. In the fourth chapter of Mark, Jesus tells of the farmer who sowed grain and watched it fall in various places, later yielding various amount of growth. The disciples did not understand the truths about the kingdom of God and asked for more clarification. Jesus continued, using the planting of seed to represent the understanding of the message of the kingdom of God.

The kingdom of God is as if someone would scatter seed on the ground, and would sleep and rise night and day, and the seed would sprout and grow, he does not know how. The earth produces of itself, first the stalk, then the head, then the full grain in the head. But when the grain is ripe, at once he goes in with his sickle, because the harvest has come. (Mark 4:26-29)

Jesus was renowned for using the natural world for illustrations in teaching the people.

God said, "Do not damage the earth or the sea or the trees" (Rev 7:3). Our planet grows weary as we pollute, waste, and ignore our neighbor's needs. God has to run on a fast track to save us from ourselves. We should raise the ceiling, not lower the floor.

Plants and Trees of the Forest

The forest is our friend, a shelter from the storms of life. It is God's spectacular and remarkable and good creation. One feels euphoric reacting to the "sound" of quietness or the chirping song of a sparrow or the sight of a deer elegantly leaping through the green meadows. The forest is a place of solitude, decorated with wildflowers such as primroses, wild violets, and buttercups that unfold graciously to reveal splendor of every shade of a palette. We must slow down or we'll miss these miracles of nature. We cannot lose these acts of creation.

The entire world is experiencing ecological damage, and eventually we may be void of the splendor of the forest. Having encountered a multitude of environmental concerns in many countries, I want to share some of my observations regarding the forest and land in Australia and in the Amazon Basin. The Amazon Basin has particular relevance for what is happening in the United States regarding the atmosphere and the ruin of our forests.

When visiting in Australia, I learned of a ecological devastation that exists particularly in the southwestern area of the country. Australia, a country boasting exotic animals, gorgeous seashores, and

South Carolina Botanical Gardens. Clemson, South Carolina.

friendly people, is facing a crisis: large portions of its western land contain a high level of saline. As much as five to ten thousand tons of salt lie under every hectare of land in these areas. Originally, the root systems of trees, which penetrate the earth some thirty to forty meters, and the shrubs and grass absorbed rainfall to keep the salt underground. However, when cattlemen arrived more than seventy-five years ago, they gradually began clearing the land for cattle and crops, causing the saline solution to rise to the surface. Now, without trees and shrubs to absorb the moisture, the rain soaks into the ground, mixing with the salt and causing excessive salinity on the land and runoffs into rivers and streams. This process has continued as more land is cleared of native plants. Not only do the few remaining trees and shrubs suffer from the salt level, but the rivers and streams are polluted as well, causing the deaths of fish, birds, and other animals that depend on the water. It is predicted that whole plant communities will die. Much of this cleared land is useless now and has been abandoned. It is estimated that it will take at least a hundred years to turn the saline tide of this land.

Another national ecological crisis that affects most countries of the world is the deforestation of the rainforest in the Amazon Basin of Brazil. I flew to Rio de Janeiro, accompanied by an interpreter, and saw evidence of the gradual disappearance of much of the Amazon Basin rainforest, which is responsible for producing 20 percent of the world's oxygen. It has been referred to as the "Lungs of our Planet" because it continuously recycles carbon dioxide into oxygen. The Amazon Basin extends more than 2.6 million square miles through seven countries with about 70 percent of the Basin in Brazil.

Prior to this trip, I had in 1989 boarded a small straw-covered boat to travel hundreds of miles down the Amazon River and viewed firsthand the exquisiteness of the rainforest. I felt I had come to the edge of the world as I saw animals on the prowl, heard the songs of multicolored birds, and encountered the wonders of the jungle. The forest floor was carpeted with flourishing green plants as well as some plants that thrived in near darkness under the canopy of virgin trees. It is a majesty to behold, even in its weathered glory. The feeling one has is inexplicable, like that sacred feeling of being in the presence of God.

Now, almost twenty years later, about 20 percent of the rainforest has been deforested. In fact, between the years of 2000 and 2005, Brazil lost more than 50,000 square miles of rainforest. In addition, the land has been cleared by fire, resulting in Brazil being a leading contributor to greenhouse gases. With this rate of clear-cutting, the forest ecology is seriously imperiled.

Then there are the loggers who have built 1,500 miles of unauthorized or illegal roads in the Amazon Basin in order to secure mahogany and other hardwoods, which they sell to the United States, Europe, and Asia. The roads cause erosion and open the area to settlement by squatters, which in turn causes more clearing of the forest and additional erosion. The runoff from the erosion goes into the rivers and creeks, causing the water supply to be polluted and filled with mud.

In addition, hundreds of people have died in land wars in the Brazilian Basin, and others live in fear or move as a consequence of fraudulent landholders and tree grabbers, who derive ludicrous profits from their illegal pursuits. Many of the 170 indigenous tribes in the Amazon Basin are gradually being driven from their land because of the new landholders (squatters) and oil drillers, many of whom are ruthless. The greed for timber and land has overridden most ecological considerations.[3]

Perhaps it would be prudent to consider a verse from Proverbs regarding those who take advantage of the poor: "Those who oppress the poor insult their Maker, but those who are kind to the needy honor him" (14:31). I visited with some of these tribes who live in one-room houses built on 10-foot-high stakes due to the flooding of the Amazon River. As I visited with one of these families, the father arrived with a 3-to-4-foot catfish, which would feed his family of six for several days. These people depend on the land and their catch from the Amazon River and its tributaries. However, as the result of the large landowners using pesticides and herbicides, people in the communities complain of "poisoned water" and dying fish. The Brazilian government is making attempts to protect the environment, but their protection agency is underfunded and understaffed, and most attempts to control the "land grabbers" in their illegal pursuits are to no avail.

The Amazon Basin issue is complex. Many of the people who live in the rainforest may be sympathetic to preserving the pristine land, but they need to feed their families. Selling their land and trees means food for the family. Richard Friedman, in his book *The Lexus and the Olive Tree*, states, "If we desire to save the land then jobs for the poor must be addressed. Save a society and you will save a tree." Brazil is only one illustration of the world's disappearing rainforests. I have also observed deforestation in Puerto Rico, Guatemala, Ecuador, and India. Throughout the world, rainforests are being destroyed due to the shortsightedness of governments that allow the removal of timber and permit landowners to "slash and burn." Friedman goes on to say that "Half of the world's tropical and temperate forests are now gone. The rate of deforestation in the tropics contiues at about an acre a second. About half the wetlands and a third of the mangroves are gone."[4]

Additionally, nearly half of the world's species of plants, animals and microorganisms will be destroyed or severely threatened over the next twenty-five years due to deforestation. Experts estimate that we are losing 137 plant, animal, and insect species every day. Currently 121 prescription drugs are derived from plants, and 25 percent of these plants are grown in the rainforest.[5]

According to Leslie Taylor in *The Healing Power of Rainforest Herbs*, "The rainforest provides innumerable sustainable products in addition to plants used for medicinal purposes. Such products as nuts, countless varieties of fruit, oil-producing plants, rubber plants, chocolate, and chicle (used for making chewing gum) are abundant." She further states that these resources—not the trees—are the true wealth of the rainforest. Experts in this area agree that leaving the rainforest intact offers more economical value than logging, clear-cutting, and burning for the purpose of growing crops. People need to help solve deforestation by creating a consumer market for sustainable rainforest products.[6]

Deforestation is a major concern regarding the preservation of the earth. The deforestation of land in the United States has mostly been on the rise since 1963, due primarily to the loss of old-growth forest land and the development of urban subdivisions. In addition, pollutants are killing forests, as are logging, the subsequent building of roads, mining, and natural disasters such as hurricanes and wildfires. Usually with deforestation there is a degrading of the environment and/or destruction of biodiversity. In 1963 there were approximately 762 million areas of forestland in the United States, but by 2050, there is a projected loss of 23 million acres if deforestation continues at the present rate.[7]

Deforestation is also caused by overconsumption of products made from trees, uncontrolled industrialization, pollution from acid rain, cattle grazing, and agriculture. Also playing major roles are inadequate research relating to such areas as the prevention of disease in trees, erosion control, pressure by trade and globalization, insufficient policies by the government, greed, and lack of enforcement of the laws that now exist regarding the environment.

On the international level, there are additional problems related to deforestation, such as wars, poverty, burning wood for fuel, clear-cutting, and illegal cutting. Thomas L. Friedman, in *Hot, Flat, and Crowded*, states that Indonesia has the fastest rate of deforestation in the world and is losing tropical forest the size of Maryland every year. He further states that the carbon released by the cutting and clearing

Deforestation is a main cause of extinction of living organisms. Seneca, South Carolina.

of their trees has made Indonesia the third-largest source of greenhouse gas emissions in the world.[8]

Perhaps people's lack of awareness is another major reason for deforestation. When a shade tree hampers one's flower garden or the growing of healthy grass, sometimes the tree is cut down without much thought as to the consequences for the environment. Developers of subdivisions thrive on clear-cutting, often leaving hundreds of acres void of trees. Perhaps we have taught an unbalanced message in our churches by neglecting the great commandment to "love your neighbor as yourself" or "do unto others as you would have them do to you." Our society seems to thrive on the philosophy that "the winner takes all," which further divides the rich and poor. It is significant to note that every tree absorbs carbon dioxide and provides oxygen. Whether trees are in the forest or in the yard, they provide sustenance for those that live below them—humans, animals, and plants. In Isaiah 55:8 the Lord scolds, "For my thoughts are not your thoughts, neither are your ways my ways." Maybe we need to pay more attention to Scriptures than to our often-flawed judgment. The environment is God's creation but our responsibility.

Our lives would be considerably hampered without the use of wood to make so many of the items we deem essential to our existence, especially paper. However, to be good green stewards, we could determine if the wood products have come from a forest that has been managed in an ecologically and socially acceptable manner. One way this can be determined is by looking for an FSC label on the item (Forest Stewardship Council, www.fsus.org). The FSC is an international nonprofit association that puts its stamp of approval on ecologically well-managed forest items. This organization also takes into consideration the effect of logging on indigenous populations that use forests for their livelihood.

For example, when I was in South Africa, I observed women and children who walked for miles to gather sticks and wood for fuel. Without a source of wood, they would be unable to cook meals for their families. Another time, while several of us were riding through jungle areas in Swaziland, our van experienced mechanical difficulty. As we got out of the vehicle, I expressed concern that wild animals might attack us. Missionaries told me that there were no wild animals in this area; the indigenous people had eaten the wild animals. The forest had for many years provided a habitat for the animals that nourished many of the tribal people. The preponderance of deforestation is happening near some of the indigenous people who suffer most, such as those in Brazil (see Deut 15:7-8).

The search for oil and gas, especially in some of the western United States, is also problematic as big industry invades rural land, building roads and destroying the habitats of wild animals. John and Teresa Kerry, in their book *This Moment on Earth*,[9] have addressed the impact of such devastation to the pristine forest in the United States. Though Kerry acknowledges that extracting fossil fuel is a "necessary part of our energy mix," he says "we . . . need to take into account its true cost and the damage it will cost." He discusses the exploration for gas in the Powder River Basin, 14 million acres of rugged prairie in northeastern Wyoming and portions of Montana. The Federal Bureau of Land Management approved a plan for natural gas development for 51,000 new wells being placed over 8 million acres of the basin, making it one of the largest gas production sites in the United States.

Children in the District of Uttar Pradesh, India, gather leaves and sticks for fuel.

Kerry further states that there are plans to explore for natural gas and oil in the Rocky Mountain West, which is one of the remaining pristine places in the country. Safeguards have, according to Kerry, been largely ignored or removed, leaving wildlife at risk. He notes that the Wilderness Society found that, during the winter of 2002–2003, the Bureau of Land Management received 172 requests from energy companies for exemptions from endangered species protection. All but three were approved.

In January 2004, President Bush announced plans to drill more than 1.2 million acres of land in Otero Mesa, New Mexico, for oil and development, which will alter forever this diverse eco-region. It is a wildlife habitat, one of the largest desert grasslands on public lands. Also, according to Kerry, energy companies are seeking to drill in the Valle Vidal, part of the 100,000 acres of the Carson National Forest in New Mexico. In the Rocky Mountain West, 95 percent of lands managed by the Bureau of Land Management are now available for oil and gas leasing. Many environmentalists and everyday citizens foresee unrecoverable losses over the "drill first and think later" approach.[10] As Proverbs 18:1 reminds us, "The selfish man quarrels against every sound principle of conduct by demanding his own way" (LB).

Portions of our national forests are open for oil and gas exploration.
Yellowstone National Park.

Is it possible that the United States could soon be seeing "For Sale" signs on large tracks of public land, including national parks? Development could destroy the wilderness and its habitat for animals, as well as mar the land with oilrigs, logging, mining, and housing developments. The Bush Administration proposed the selling of 800,000 acres of public land in 35 states, of which 300,000 acres are national forest land. It includes portions of the Ocala National Forest in Florida, the Columbia River Gorge National Scenic Area in Oregon and Washington, the DeSoto National Forest in Mississippi, and the Klamath National Forest in California. If these and even more recent proposals materialize, we will lose this public land forever, according to the Wilderness Society. [11]

Another factor that has tremendously affected forests in the United States is acid rain. More than twenty years ago, our family traveled to the Great Smoky Mountains National Forest to observe the spectacular view of lush greenery and mountains and an occasional mother bear and her cubs. As we approached some of the higher elevations, we observed a few areas with dead trees. As we arrived at the peak of the mountains, we found that most of the trees were not merely dying but dead. We learned that one of the contributing factors to their demise was acid rain. Earlier, when visiting my mother in Mississippi, I walked over to her neighbor's yard to see her small goldfish pond, but there were no fish. I was dismayed to discover that the goldfish had died as a result of acid rain. Before these experiences, I had heard of acid rain but had not observed the results firsthand.

Acid rain forms when sulfur dioxide and nitrogen oxides combine with moisture such as clouds, rain, sleet, fog, snow, or mist. These fossil fuels are dispensed into the air from factories, power stations, cars, and even homes. When acid rain is accompanied by stress factors such as air pollutants, insects, disease, drought, or very cold weather, trees die. Foliage is also weakened as the result of the soil's absorbing acidic rain, especially in the eastern United States. When leaves and other greenery are damaged by acid rain, they are resistant to absorbing nutrients, and photosynthesis that provides their food for growth is inhibited. Trees in high mountain regions are often exposed to acid rainfall as a result of being surrounded by acidic clouds and fog.[12]

Other changes caused by acid rain include the reduction of red spruce and sugar maple trees, changes in the chemistry of oceans (which cause the reduction of salmon, especially in the Atlantic), and the demise of fish in streams and lakes. Food crops, however, are less at risk from acid rain because farmers restore soil conditions by adding fertilizers to replace lost nutrients. Fertilizers, of course, present their own range of problems.

Acid rain also damages historic buildings and monuments, such as the Statue of Liberty, which required restoration due largely to acid rain. Buildings erode from natural elements, such as sun, rain, and wind, but acidic gases accelerate the erosion. Scientists believe that if we lower the level of gases, we could significantly reduce acid rain.

Another concern that plagues the United States is that of non-native plants that overwhelm the forests. In fact, anywhere there is land, we can find these persistent and insidious plants. Kudzu is one such example. Native to Asia, it was introduced to the United States in the late 1800s as an ornamental and for erosion control.[13] Several years ago in our neighborhood, a wooded lot needed to be cleared for the building of a new home, but before the builder could pour the footings for the house, he had to remove the kudzu. This "vine that ate the South" was smothering the native trees on the lot and providing a lush green coverage for the ground. The stems of the vines, some the size of a human arm, were removed by attaching a chain from a tractor to the kudzu and pulling the vine from the trees. However, the roots and remaining kudzu vines gave life to new outcroppings of kudzu, requiring years to control since it can grow up to 60 feet a year. In fact, ten years later the kudzu is still "king," surrounding the edges of the yard of this home. Such is often the plight in the southern and eastern U.S. The climate is perfect for supporting the already existing 7 million acres of kudzu, which does prevent erosion but has few other uses. Channing Cope, a reporter for the *Atlanta Journal*, writes that "Cotton isn't king here anymore. . . . Kudzu is king!"[14] James Dickey, in his poem "Kudzu," expresses many of our sentiments, writing, "you must close your windows / At night to keep it out of the house."[15]

Non-native plants and trees pose serious problems because they inhibit sunlight to plants and destroy valuable native forests. Kudzu is

Kudzu is now "king of the South." Seneca, South Carolina.

only one of more than 97 biological invasions of non-native plants (whether they be trees, vines, shrubs, grass, ferns, or aquatic plants[16]) in the South and Southeast. These invasive plants are one of the leading causes for the loss of native species. Non-native plants also disrupt the habitat of animals, resulting in the loss of birds and insects. It is an ecological problem that has gained minimum publicity but desperately needs to be addressed.

Many of these non-native plants have been brought to the United States for ornamental purposes or as forage for livestock. Americans must participate in the eradication of these plants through concerted control measures with USDA-approved herbicide sprays, stem injections (placing herbicides in an incision on the tree trunk), cut-treatment (cutting the tree and placing herbicide on the stump), or simply by pulling up these plants as they appear in one's own backyard.[17]

Others invasive plants include Japanese climbing ferns, garlic mustard, Chinese lespedeza, tall fescue, Johnson grass, Chinaberry tree, English ivy, common periwinkle, Norway maple, Japanese barberry, autumn olive, yellow iris, multi-flora rose, St. Johnswort, Japanese honeysuckle, Russian olive, and giant reed. These plants arrived without their native predators of insects and diseases, which kept the plants in natural balance within their native habitat.

Some native and non-native plants are growing faster and becoming more potent with the warming of the climate. Most of us have encountered the three-leaf plant that every child is warned to avoid—poison ivy, whose vines wrap around trees and bushes in the woods and in backyards. Just touching the vine can cause itchy, irritating blisters that last for days. Now we learn from researchers at Duke University that poison ivy grows bigger and faster as carbon dioxide levels increase in the atmosphere. High levels of carbon dioxide produce more of the rash-causing chemical, urushiol, an oil contained in the poison ivy. More than 350,000 cases are reported each year, according to poison control centers.[18]

The Land and the Marshes

Soil is vital to the survival of humans and animals, providing both food and habitat. It possesses nutrients and water that are essential for growing trees, plants, vegetables, and grasses. It is the foundation for our highways, homes, businesses, schools, and hospitals, and is vital to our recreation and way of life. Not only do we glean resources from the soil, but it also helps to protect us from the aquatic ecosystems.

Since land is so vital to us, why do we contaminate our soil with such abandon? There is urban sprawl, excessive drainage, clear-cutting of trees, loss of habitat, erosion, overgrazing, and industrial waste, some accidentally and others deliberately. Prior to the passage of laws prohibiting dumping of industrial waste, some land was contaminated. The contamination remains in the soil unless there is a concerted effort to remove it. Dumping refuse continues to be a serious problem. Land is often used as unauthorized garbage dumps, especially in rural areas, while most cities dispose of waste according to regulations. Recently as we traveled in Mississippi, I saw areas along the highways where it appeared that residents simply had dumped their weekly garbage. Two years ago, within a mile of our residence, a sign had to be installed along a state road that read, "No dumping." These dumping sites are hazardous to communities and injure the land.

Twenty years ago, we owned a farm about twenty-five miles from our residence in the foothills of South Carolina. This farm began as my husband's hobby, but it developed into a business. He maintained about 2,000 pigs on the farm, so pollution became a major concern. He adhered strictly to state regulations regarding lagoons, such as runoffs and maintaining the lagoon a suitable distance from streams. When the farm was acquired, it was located in an isolated rural area, but eventually new houses were built within a half-mile, and the odor from the lagoon became a problem. He eventually sold all his pigs and closed the farm. Maintaining an odor-free, pollution-free area with several thousand hogs was a challenge. In eastern North Carolina, hundreds of large hog farms face major pollution challenges resulting from waste that leaks into the soil and streams.

Land use is not a black-and-white issue. While humans need food and natural resources, safeguards against the misuse of land and its resources are essential. Developing subdivisions and roads, changing the channels of streams and damming rivers, and removing land cover are often considered signs of progress, but we need to be more discriminating in how we manage these endeavors. Counties, states, and the federal government must develop and enforce regulations to provide safety and well-being, not only for the present but also for the future.

Cultivating mangroves is another essential part of the ecosystem that deserves careful management. Mangrove swamps are found in tropical and subtropical regions over the world. In North America, they extend from Florida along the Gulf Coast to Texas. The coast of southwest Florida has one of the largest mangrove swamps in the world. These swamps contain trees, shrubs, and others plants that grow in brackish to saline tidal waters. These mangroves provide protection from shoreline erosion, especially during hurricanes and tidal waves. When I was a student in New Orleans, I often traveled by small boat into the mangrove areas along the southern tip of Louisiana. Now many of these areas have been filled in for developments and the trees removed, disrupting the ecosystem. It is believed that one reason New Orleans was so devastated during Hurricane Katrina was the loss of mangrove swamps. Not only do the mangroves protect land from

Mangroves are essential to the ecosystem. Near the Great Barrier Reef, Australia.

water damage along the coastline, but they are also home to shrimp, oysters, worms, protozoa, and other invertebrates. In turn, these organisms feed crocodiles and wading birds like pelicans. These swamps "are constantly replenished with nutrients transported by fresh water runoff from the land and flushed by the ebb and flow of the tides."[19]

Land, flora, and fauna suffer from wildfires. Recently our local South Carolina newspaper headline read "Smoke Chokes Upstate."[20] DHEC warned that the local air quality was unhealthy, producing dangerous ozone levels for sensitive groups as the result of the winds fanning smoke northward from the Georgia and Florida wildfires. There is a wealth of evidence that these fires that menaced the South and the West resulted from drought, another possible effect of global warming.

Erosion

I cautiously walked down convoluted streets of dirt, mud, and broken pavement that twisted through the crumbling city of Port-au-Prince, Haiti. Crowds of dark-skinned people displayed their wares on the ground or in impromptu storefronts on either side, children played in the street, and old women sat quietly. A few people washed themselves from an open spigot, with the wastewater flowing down the streets, creating erosion and exposure to all types of germs and disease. The stench from the open sewage flowing down the middle of the streets and the garbage, piled six feet high, made me wonder how these people survived.

The erosion in the city and in the countryside presents one of the most devastating plights to a people that I have observed. Haiti is the poorest country in the Western hemisphere, with more than 8 million people living in squalor. Almost every inch in the city is occupied, mostly by lean-to houses but with a few homes for the well-to-do, political leaders, and merchants. In the countryside, all the useful land is cultivated to grow vegetable crops or sisal for making baskets. Due to lack of knowledge about terracing and erosion prevention, hillside cultivation erodes the land and washes pollution into streams and the sea. Trees were removed primarily for firewood and to provide land for growing crops, creating additional erosion.

As I saw the plight of the Haitian people and their need for an improved way of life, more than ever I realized that as Christians we

More than two billion tons of cropland soil erosion occurs yearly. Port-au-Prince, Haiti.

have an obligation to minister to the dilemma of these people. I recalled Jesus' words in Matthew 25:40: "Truly I tell you, just as you did it to one of the least of these who are members of my family, you did it to me." It is important that we be concerned regarding erosion. It destroys land and the resulting runoff becomes a pollutant as it flows into rivers and streams. In many developing countries, such as Ecuador, new land is cleared every few years because of the lack of fertilizers and other ingredients to enhance the soil for growing crops. The people "slash and burn" new wooded areas to secure fresh land for their crops and abandon the old land to erosion and neglect. Once soil is damaged or contaminated, it is difficult to restore to its natural fertile status.

Wind and water erosion resulted in 1.9 billion tons of cropland soil erosion in 1997, according to the Natural Resources Conservation report.[21] A federal conservation program was established under the Food Security Act of 1985 to assist landowners in providing ground cover to prevent the eroding of the land. Natural erosion, which occurs over long periods, is referred to as "geological erosion," while "accelerated erosion" refers to erosion caused by human activity. Gully erosion happens when a narrow stream erodes, washing soil from banks and causing gullies as deep as 100 feet. Erosion follows heavy rain, especially during the growing season when there is limited coverage of the land. Wind erosion occurs when wind blows the soil into other areas.

Another injury to the earth is destruction due to war. I was in Kosovo only three months after the Serbian-Kosovar war (1999) and observed the ruins to God's good earth—land mines, burned-out villages and forests, destruction to crops in the fields, dead cattle, poisoned wells, and the general disruption of the flow of nature. Second Samuel 2:26 reads, "Must our swords continue to kill each other forever. . . , " and my experience in Kosovo made plain that the destruction of war lasts far beyond the actual fighting. Tanks and other military vehicles had made deep ruts in the dirt roads, creating muddy passages for residents and travelers. With ten to fifteen nations at war at any given time, one realizes that destruction is inevitable. We monitor daily the bombings in Iraq and flinch at not only the wounded

The ruins of war. Begrac, Kosovo.

and deceased but also at the damage to the surroundings. War is one of the environmental ravages that fortunately we have not encountered on a large scale in the United States.

I did not have to go to Haiti or Kosovo to observe pollution or erosion. Just behind our house, following a heavy rain, a muddy lake pools as a result of loose soil that enters it from nearby feeder streams. The mud is apparently residue from new housing developments and other construction work. Workers often ignore state requirements for building silt fences and other erosion control measures.

Agriculture

In Genesis, God tells his human creation, "See, I have given you every plant yielding seed that is upon the face of all the earth and every tree with seed in its fruit; you shall have them for food" (Gen 1:28). What we eat affects the environment. On average, produce travels about 1,500 miles from farm to supermarket, according to Sustainable Table, a service of the Global Resources Action Center for the Environment.[22] Trucks that transport food are tremendous polluters as they travel back and forth from large agriculture areas, such as

California and Florida, to cities throughout the United States. In addition, approximately 40 percent of the fruit we consume is produced outside the United States and has to be transported from seaports to our local markets. By purchasing locally produced food or asking the grocer to purchase foodstuffs grown by local farmers or at least grown within our state, we could eliminate some of the cost of transportation and in turn reduce the use and cost of gas and oil. Homegrown foods are available at local farmers' markets and are usually fresher (and cheaper!). By purchasing locally produced products, we benefit the economy of our communities—clearly a win-win situation.

Many individuals are choosing to purchase organically grown foods to avoid pesticides, chemical fertilizers, and additives. According to the Mayo Clinic Research,[23] the benefit of purchasing organic foods is the enhancement of the environment through eliminating conventional fertilizers, encouraging soil and water conservation, and reducing pollution. Organic farmers use only natural fertilizers such as manure and/or compost, and the products have about the same appearance as non-organic foods. With organic meat, the food is derived from animals that have been fed only organic feed, have access

Aloe trees provide a livelihood to farmers in Aruba.

to free range (the outdoors), and have not been administered antibiotics, growth hormones, or other medications to prevent diseases.

When buying food, look for a small green sticker with "USDA-ORGANIC" to determine if the product is organic. If the product is "100 percent organic," it was grown under the best conditions as stated above; if "organic," the food has at least 95 percent organically produced ingredients. "Made with organic ingredients" indicates that the food has at least 70 percent organic ingredients. It would also be expedient to determine where the produce has been grown because the United States has more stringent regulations and inspections than do most other countries.[24] Perhaps the main hindrance to increased purchase of organic foods is that they are more expensive.

Recently I read a copy of my husband's weekly agriculture newspaper, *Feedstuffs,* and was interested to find three articles related to environmental issues. One of the articles was "Biotech Making Farming Greener." Biotechnology is a scientific method that improves or modifies a plant's genetic structure. Most crops grown for food utilize some form of genetic modification, a technique used for centuries to gain the best seeds, produce plants resistant to disease, and develop characteristics such as drought resistance. Genetic modification allows the selecting of genes that produce desired traits, a process beneficial both to farmers and the environment. This study found that biotech crops increase yield and allow farmers to practice "conservation-focused low- or no-till practices in their farming." This process reduces fieldwork and pesticide use and lowers farm fuel consumption, thus providing a lower production cost, with obvious benefits to the ecology and the reduction of carbon dioxide emissions. This study also stated, "in 2005 biotech crop production reduced carbon dioxide emissions by nine billion kilograms which is equivalent to taking four million cars off the road for one year."[25]

Another article in the same issue of *Feedstuffs* dealt with research at Penn State's College of Agricultural Sciences regarding the disposing of poultry litter (wood shavings and manure) in a way that produces energy and reduces water pollution at the same time. Because poultry growers spend enormous amounts of money to heat their chicken houses, they are attempting to convert the litter to fuel. One of the researchers stated that burning one pound of poultry litter produces about a third of the energy from burning one pound of coal. This fuel, in turn, could be utilized to heat the chicken houses and eliminate manure runoff that contaminates land and surface water. Burning the litter as fuel produces a phosphorus-rich ash, which could be bagged and sold as a plant fertilizer. How encouraging to know that innovative businesses are considering environmental benefits as they consider ways to improve their process.[26]

Earth Justice

According to the April 2007 *Newsweek,* there seems to be a greater acceptance by skeptics regarding climatic changes and the inevitable consequences. To overcome prevailing cynicism on the global dilemma, the Intergovernmental Panel on Climate Change released a statement announcing research by 1,000 scientists from seventy-four countries. Their report stated that "global warming was already affecting the Earth's ecosystems . . . that continued climate change, in combination with other environmental stressors such as population increases and greater urbanization, would lead to more-severe and widespread drought, greater coastal and riverine flooding, and 'increased risk of extinction' for 20 to 30 percent of plant and animal species."[27]

There seems to be a correlation between the preponderance of evidence regarding climatic changes and the acceptance and concern by the public. An enlightened people will more likely react to necessary measures and address these vital issues. Most of those who have opposed pollution controls have felt their interests were threatened, considered the technology required to change insurmountable, and deemed the changes too costly. Industries balk at controlling carbon dioxide emissions, but there is at least some change of heart toward renewable fuels such as ethanol. A limited number of companies are moving toward sustainability, realizing that resources are indeed limited.

Are we being "green stewards" when we seek to use the world's resources to promote the affluence to which Americans are addicted?

Are we willing to make changes? Are we willing to spend a little more for energy-saving light bulbs? Are we willing to car pool, walk more, and ride bicycles to save on gasoline? Or are we ethically lethargic regarding the environment? Americans consume 25 percent of the world's resources, yet we make up only 2 percent of the earth's people.

Could we promote "justice for all" by planting trees to help transform the environment and our quality of life? Perhaps this is one way we could leave our ecological footprints on the earth. Trees can help remove air pollution by lowering air temperature. Without trees, sunbaked countrysides and cities experience more blowing soil, silty flood streams and rivers, and erosion from water runoff. Trees can protect fields, enhance natural streams, provide homes for birds and nesting sites and food for a variety of wildlife, and offer cooling shade for homes and cities. The Arbor Day Foundation Tree City USA program, one of the oldest environmental causes in the nation, says that 100 million mature trees in United States cities would save 4 billion dollars per year in energy. They suggest that more trees increase property values by shading concrete driveways, cut cooling costs by 15–35 percent, and provide a windbreak that can reduce heating costs. Often we are frustrated by the enormity of major projects, but we can begin by doing our part and planting one tree at a time.

Can we as Christians experience and communicate with God in the same framework as before when we observe the degradation imposed on his creation right in front of our eyes? Do we need a spiritual renewal to live simply by avoiding excessive consumerism and defining our priorities? Martin Luther King said, "Our lives begin to end the day we become silent about the things that matter."

Questions for Reflection

1. How does the command to "love your neighbor as yourself" give rise to concerns for the environment?
2. How do we sometimes miss the miracles of nature in our hectic lives?
3. Indigenous people of the Amazon Basin depend heavily on harvesting timber and "slash-and-burn" methods in clearing land for their

livelihood, yet a portion of the world's oxygen is being eliminated in the process of demolishing the Amazon Forest. What is the solution to such a situation?

4. How does consumerism hamper creation care?

5. How can our relationship with Christ make us more conscientious regarding creation care?

6. Discuss the role of greed in relationship to deforestation, pollution, acid rain, and erosion. What are some ways to resolve these difficulties?

7. How do our lifestyles perpetuate negativity and apathy toward the environment?

8. Why do we revere God's word but sometimes fail to heed the message of caring about the environment and our neighbors' welfare?

Notes

[1] Rachel Carson, *Silent Spring*, 40th anniversary ed. (New York: Houghton Mifflin Company, 2002), 6.

[2] Parham's article on Ethicsdaily.com is no longer available, but there is a great deal of useful information on the Green Bible at the Ethicsdaily.com Website: http://www.thegreenbible.org.

[3] Scott Wallace, "Last of the Amazon," *National Geographic* 211/1 (January 2007): 40–71.

[4] Thomas L. Friedman, *The Lexus and the Olive Tree: Understanding Globalization* (New York: Anchor Books, Random House, 2000), 280, 46.

[5] "The Disappearing Rainforests," Rainforest Facts, www.rain-tree.com/facts.htm (May 2006).

[6] Leslie Taylor, "The Importance of the Rainforest," in *The Healing Power of Rain Forest Herbs* (Garden City NY: Square One Publishing, Inc., 2004).

[7] "Deforestation, United States," http://en.wikipedia.org/wiki/Deforestation #United_States (30 December 2008).

[8] Thomas L. Friedman, *Hot, Flat, and Crowded* (New York: Farrar, Straus & Giroux, 2008), 299.

[9] John Kerry and Teresa Heinz Kerry, *This Moment on Earth: Today's New Environmentalists and Their Vision for the Future* (New York: Public Affairs, 2007), 150–53.

[10] Ibid., 154.

[11] "A Bad Deal for America's Wild Lands: Regulation Change and Environmental Rollbacks in the Bush Administration's Waning Days," The Wilderness Society, http://wilderness.org/content/bad-deal (26 November 2008; accessed 30 December 2008).

[12] Ibid.

[13] "Kudzu," National Invasive Species Information Center, United States Department of Agriculture, http://www.invasivespeciesinfo.gov/plants/kudzu.shtml (4 December 2008).

[14] C. Ritchie Bell, and Charles Reagan Wilson, "The Kudzu File," from *The Encyclopedia of Southern Culture*, ed. Charles Reagan Wilson and William Ferris (Chapel Hill: University of North Carolina Press, 1989).

[15] Ibid.

[16] Charles T. Bargeron, et al., "Invasive Plants of the Eastern United States: Identification and Control," The University of Georgia, USDA APHIS PPQ and USDA Forest Service, Forest Health Technology Enterprise Team, http://www.invasive.org/eastern/ (5 December 2008).

[17] James Miller, *Normative Invasive Plants of Southern Forest: A Field Guide for Identification and Control* (Asheville NC: U.S. Department of Agriculture, Southern Research Station, 2003), 93.

[18] Sierra Club Newsletter, September–October 2006, 16.

[19] "Mangrove Swamps," Environment Protection Agency, www.epa.gov/owow/wetland/types/mangrove.html (February 2006).

[20] *Greenville* (SC) *News*, 24 May 2007.

[21] "Total Wind and Water Erosion," 1997, Natural Resources Conservation Service, http://www.nrcs.usda.gov/technical/NRI/maps/meta/m5112.html (30 December 2008).

[22] "NEWS: Could You Go Locavore?" Sustainable Table, http://www.sustainabletable.org/blog/archives/2005/06/index.html (29 June 2005; accessed 30 December 2008).

[23] "Organic Foods: Are They Safer? More Nutritious?" MayoClinic.com, http://www.mayoclinic.com/health/organic-food/NU00255 (30 December 2008).

[24] "Buying Organically Grown Food," University of Minnesota Extension, http://www.extension.umn.edu/info-u/nutrition/bj869.html (30 December 2008).

[25] Rod Smith, "Biotech Makes Farming Greener," *Feedstuffs* (7 May 2007).

[26] Ibid.

[27] Jerry Adler, "Moment of Truth," *Newsweek*, 16 April 2007, http://www.newsweek.com/id/35655 (5 December 2008).

Chapter 4

THIS LAND OF OURS: THE FAUNA

My knees rested on the unyielding cobblestone floor as I bowed my head to offer thanks at the birthplace of Saint Francis of Assisi. Here, in a small Italian village, this Franciscan monk was born into a prosperous family, but in 1201 he renounced all material possessions in order to follow God. Saint Francis was renowned for his brilliant mind, his obedience to God, and his devotion to animals and nature.

I first became acquainted with the contributions of Saint Francis to the animal kingdom through my pastor, Rev. Charles Arrington, more than thirty years ago. On Sunday mornings during children's story time, Rev. Arrington often would tell stories of Saint Francis's care of animals or his deeds to benefit mankind. These stories, plus numerous books about Saint Francis, have continued to intrigue me to the point that, when I was traveling in Italy, visiting Assisi was a top priority.

When a close friend and I finished three weeks of mission work in Albania in 1996, we flew to Rome and then left by bus for Assisi. We traveled through the rolling hillsides and up the slopes of Monte Subasio, Spello, and Treni to arrive at the walled village of Assisi. The region was pristine, with vineyards and olive groves dotting the landscape between villages. Houses, painted an array of pastel colors and nestled along narrow streets, were adorned with flower boxes laden with dazzling summer flowers. My impression was that the people took great pride in caring for the earth.

Prior to the time of my visit to Saint Francis of Assisi's home and church, an earthquake had destroyed portions of the Church of Saint Mary of Angels, although Francis's humble home, only yards away, had been spared. Saint Mary's, a magnificent edifice, was filled with mar-

Saint Frances of Assisi proctecting the birds in our yard in Seneca, South Carolina.

velous paintings and icons as well as the original chapel where Saint Francis and other monks had worshiped. The chapel was small, perhaps twenty by thirty feet, and had been preserved within Saint Mary's. I was filled with awe and gratitude as I participated in a Catholic mass in the chapel where Saint Francis had worshiped until his death in 1224 at the age of forty-four.

Much has been written regarding the sainthood of Saint Francis, but his unique ability to communicate with animals may be unsurpassed. From Fr. Johannes M. Boali's book, *The Little Flowers of St. Francis*,[1] come intriguing stories, and one story titled "How St. Francis Tamed the Wild Turtle-Doves" is typical.

One day Saint Francis approached a boy carrying turtledoves to sell at the market:

Whoever had singular compassion for gentle creatures, chanced to meet him, and looking upon those turtle-doves with compassionate eyes said, "Good youth, I pray thee give them to me, that birds so gentle, which in the Scriptures are likened unto chaste and humble and faithful souls, come not into the hands of cruel men who would slay them." St. Francis took the birds and held them closely speaking sweetly to them saying, "O my sisters, simple, innocent, chaste turtle-doves, why do you let yourselves be taken. Now I desire to save you from death and to make a nest for you so that ye may bring forth fruit and multiply, according to the commandments of our Creator."

So Saint Francis took the birds, made nests for them, observed their laying of eggs, and watched the hatching of young birds. These birds were as tame as domestic fowls and would not leave either Saint Francis or the friars until Saint Francis gave them his blessing and permission to leave. Saint Francis said to the young man from whom he took the birds that he should become a friar and live in the order in great sanctity, and he did.

Recognizing this great communicator with animals, Pope John Paul II said, "Saint Francis of Assisi was proclaimed as the heavenly patron of those who promote ecology. He is an example of genuine and deep respect for the integrity of creation. He was one who was loved by God's creatures and invited all of creation—animal, plants and natural forces—to give honor and praise to the Lord."[2]

God Honors Animals

Creative supremacy was often noted in God's concern for animals, as is evident in the first chapter of Genesis. God says to his human creations, "And to every beast of the earth, and to every bird of the air, and to everything that creeps on the earth, everything that has the breath of life, I have given every green plant for food. And it was so. God saw everything he had made, and indeed it was very good" (Gen 1:30-31). We too often fail to be realistic in our responsibility for the animal kingdom. God designated humans to be caretakers, indicating that we should safeguard the animal ecosystem.

Another verse indicating man's responsibility to value and respect animals is found in Deuteronomy 25:4 when God says, "You shall not muzzle an ox while it is treading out the grain." In Job, God addresses his dominion over animals:

Can you hunt the prey for the lion?
or satisfy the appetite of the young lions
when they crouch in their dens
or lie in wait in a covert?
Who provides for the raven its prey
when its young ones cry to God
And wander about for lack of food? (Job 38:39-41)

Marcia Bungee in "Biblical Views of Nature" says these verses in Job indicate that "God takes great delight in non-human creatures and did not create them for human benefit alone. Such passages all imply that human beings need to respect nature, to recognize the intrinsic value of its many creatures, to learn from it, and to preserve its incredible diversity."[3]

In addition to these verses in Job, we find Paul addressing the Corinthians regarding their eating of meat sacrificed in the temple. Paul instructs the people to say a blessing over the food; there will be no objection to eating any of God's creatures when a blessing is offered, for "the earth and it fullness are the Lord's" (1 Cor 10:26). People have freedom of consciousness, according to Paul, to make decisions for themselves; however, they should be careful not to abuse this God-given freedom.

Also God aspires for humans and animals to multiply and replenish the earth. The biblical account of Noah and the flood relates God's yearnings when he commands Noah to "Bring out with you every living thing that is with you of all flesh—birds and animals and every creeping thing that creeps on earth—so that they may abound on the earth, and be fruitful and multiply on the earth" (Gen 8:17). God then made a covenant with Noah, telling him that no living beings

The water and land need protection if the Bison are to survive.
Yellowstone National Park.

would be destroyed ever again by a flood (Gen 9:8-11). We are all connected to the earth and to each other through God's pledge to us.

In addition, the writer of Isaiah places high value on animal life as he uses animal imagery to describe God's relationship to his people. God offers words of hope, comfort, and deliverance for the people of Israel:

> He will feed his flock like a shepherd;
> > he will gather the lambs in his arms, and carry them in his bosom,
> and gently lead the mother sheep.
> > But those who wait for the LORD shall renew their strength,
> they shall mount up with wings like eagles,
> > they shall run and not be weary, they shall walk and not faint.
> (Isa 40:11, 31)

Isaiah also prophesies that the Davidic kingdom will be a peaceful kingdom and prevail: "The wolf shall live with the lamb, the leopard shall lie down with the kid, the calf and the lion and the fatling together, and a little child shall lead them. The cow and the bear shall graze, their young shall lie down together; and the lion shall eat straw like the ox" (Isa 11:6-7).

These verses of Isaiah came to mind as my husband, Woodie, and I visited the Galápagos Islands 500 miles off the coast of Ecuador. Fourteen of the nineteen islands are uninhabited and remain a pristine habitat for animals. The Ecuadorian government has established strict guidelines for maintaining the ecosystem of the islands, requiring a guide for every ten people who visit the islands. This measure ensures that tourists adhere to an identified path and avoid disturbing the environment that is the home to thousands of birds and animals. More astounding to me than finding the 600-pound, 200-year-old turtles or observing the penguins, iguanas, sea lions, red- and blue-footed boobies, or the dozens of varieties of birds was the docility of these animals. They appeared domesticated, undaunted by humans and fearless of each other. For example, I observed a sea lion, an iguana, and a blue-footed booby sunning together under a shade tree within inches of one another. In that these animals are unmolested by humans or other animals, theirs is a peaceful kingdom. Their kingdom is tranquil and calm just as in the prophetic description of the Davidic kingdom, in which the "lion and the lamb" lie down together. This is the hope for our world.

Blue-footed boobies inhabit the Galápagos Islands.

Endangered Species

Birds and animals have always captivated me. As a child, I fished in the creeks and ponds in Mississippi, hunted animals for food with my brother, Charles, and tended the domestic animals on our farm. Animal life was woven into every segment of my life, including the realities of poisonous snakes and butting goats. As an adult, I'm still mesmerized by God's provision of so many unique creatures, creatures such as iguanas, which are rather gruesome and remind me of dragons. I have ridden elephants in Thailand, observed black bears in Russia, watched the grazing elk and musk ox of Alaska, been charmed by the grazing zebra and gazelles in Africa, been captivated by the koalas and kangaroos of Australia, and cuddled a three-toed sloth in Peru. I have never had my fill of animals.

Exotic animals are enchanting, but I am equally taken when a whitetail deer or fox dashes across our backyard or a beaver visits our lake shoreline or when a raccoon visits our deck to pilfer the bird feed. I have found my sense of place in the world by mingling with these creatures that are provided to us for a purpose and for our benefit. God's magnificent creations remind me of the extent of his love and concern for me. It troubles me to think that my grandchildren may never see some of these creatures.

In order to protect living creatures, we must maintain their habitats—woodlands, wetlands, and forests. With large tracts of land being urbanized on every corner, there is massive loss of animal habitat. Approximately one-third of the animals and plants are in jeopardy. It is encouraging that some of the endangered species are beginning to make a comeback. This is due largely to conservation efforts by organizations such as the National Wildlife Federation, which recently reintroduced wolves to Yellowstone National Park and is now directing a major effort toward saving the endangered Florida panthers in the Everglades.

The Endangered Species Act was passed in 1973 to encourage states in developing and maintaining conservation programs to safeguard fish, wildlife, and plants. Attached to the bill was federal financial assistance whereby the ecosystem could be preserved for threatened or endangered species. The act required that appropriate

Some turtles, weighing hundreds of pounds,
have become extinct on the Galápagos Islands.

steps be taken to alleviate threats to species of animals identified as
endangered or threatened.[4] "Endangered species" indicates that the
animal is at risk for extinction, while "threatened species" signals that
the species may become endangered in the future.

However, to place an animal on the list as endangered or threatened, an individual or organization must petition the government,
and the process can take years before the species is placed on the list.
The request may be denied for a number of reasons: lack of evidence
regarding the status of the animal, insufficient funds, or influence by
politicians who affect the decision-making process. An example of a
long waiting period is the Florida Black Bear, which was a candidate
from 1992 until 1998.[5]

In the United States there are 496 species of animals classified as
threatened or endangered, while more than 1,000 animal species are
endangered worldwide.[6] The foremost reasons for extinction of animals and birds are loss of habitat, over-harvesting, and pollution,
which can result in death, disease, and sterilization. Other causes relate
to new predators and competition for food.

I observed an example of competition for food on the Galápagos
Islands: the goats are almost obliterating the grasses that are the

primary source of food for the turtles and other animals. More than 100 years ago, when sailors arrived on the islands from Europe, they brought goats so they could enjoy fresh meat. However, when the sailors left, some of the goats remained and their numbers have increased through the years. With their constant foraging of the grasses and other flora, food for the turtles has been substantially diminished. Three subspecies of the tortoise have become extinct. Efforts have been made by the Ecuadorians to eliminate some of the goat population in order to enhance the food supply for the turtles and other species on the islands; however, the goats can be difficult to find in these unexplored territories.

Estimates by the Eden Conservancy indicate that at least two species of God's creation go extinct every day, primarily as a result of man's behavior to secure short-term gains; thus, "The creation that is called to worship God is silenced."[7] Paul says, "For we know that even the things of nature, like animals and plants, suffer in sickness and death as they await" the glorious freedom found in God's return (Rom 8:22, LB). There may be no more tangible evidence of this suffering than a view of our own towns and cities where deforestation and sprawl have diminished thousands of acres of God's creation, destroying the animals' natural world. We need to search our souls to find ways to impede the assault on God's handiwork.

In South Africa, the government protects national park reserves. Here one encounters all kinds of wildlife—herds of elephants stomping everything in their path, zebras foraging for food, hippopotamuses (some species endangered) wallowing in ponds, baboons climbing atop one's car, leopards dashing through the forest, impalas leaping through the grasses, and a mother hyena nursing her baby as she meanders through the forest. I experienced a real scare on this journey when I sought just the perfect photo, finding that I had inadvertently invaded a warthog's space. Squealing and snapping its jaws, it chased me for a short distance. It was a lesson I remember well—wildlife is to be treasured from a distance. Is there a probability that these animals will become extinct as have so many others? According to an article in the August 6, 2007, *Newsweek*, 510 mammals, 174 reptiles, and 1,180 amphibians are now endangered species.[8]

Another encounter I had with wild animals was in Peru. I, along
with several university colleagues, drifted down the Amazon River for
hours, finally arriving at our destination—a camp that was to be our
headquarters for about a week. I walked through clouds of mosqui-
toes, past squawking macaws, cautiously avoiding a snake that dangled
from the rafters over the outdoor eating area, and approached my one-
room cabin that was situated well into the tropical forest. As I
sauntered down a boardwalk to my cabin, I carefully herded back into
the wetlands a small alligator that was sunning at the entrance. I was
cautioned to look over the room, under the bed, and everywhere for
poisonous red frogs, and I did so meticulously without further encour-
agement. I encountered an adventure with wildlife at every turn!

However, if I were to return today, I just might find that the red
frogs are no longer a part of this Amazon jungle area. Scientists have
stated that global warming has been linked to the extinction of as
many as sixty closely related species of frogs that have vanished in
Latin America. With the warming trend, diseases have become ram-
pant, eradicating some species of frogs.[9] NASA scientists, led by James
Hansen of the Goddard Institute of Space Studies, reported in
September 2006 that due to the rapid warming trend over the past 30
years, the earth is at its warmest level in the last 12,000 years. Some
scientists indicate the earth's average temperature rose roughly one
degree during the twentieth century.[10]

Also, polar bears face threats from industrialization, toxic chemicals, and over-hunting of some of the species. Plans are in the making
by the United States government to allow oil drilling in the Beaufort
Sea, just off Alaska's Arctic National Wildlife Refuge and the Western
Arctic Reserve, which may further deteriorate the polar bears' habitat.
The Natural Resource Defense Council and other conservation organizations support the United States Fish and Wildlife Service in seeking
to have polar bears listed as a threatened species. "By law," they say, "if
this happens, it will mean that federal agencies will be required to
ensure that their actions do not put the polar bear at risk of extinction
or harm its critical habitat."[11]

In addition to the polar bears, there are other endangered animals
in North America. Certain species of alligators, crocodiles, salaman-

A sea lion basking in the morning sun on the Galápagos Islands.

ders, sea turtles, bats, flying squirrels, prairie dogs, bighorn sheep, bison, black bears, elk, gray wolves, mountain lions, and the pronghorn antelopes are just some of those endangered. When these animals suffer, we suffer; and when they are healthy and prospering, so are we, for we are all part of creation and dependent upon one another. With the power of our minds and modern technology at our disposal, we need to be a protective—not destructive—force. The loss of animal life in the United States is staggering, and many species of animals that once roamed our forest are now extinct or threatened.

In sharp contrast to our apathy toward wild animals, Americans spend enormous amounts of money on domestic animals—for food, healthcare, and even insurance and burial policies for our beloved pets. Though we say that we are concerned about imperiled wild animals, our behavior usually does not reflect such concern, especially for the extinction of rare species.

Birds Leaving Their Nests

There are probably no wild creatures more beloved than birds. Is there anything more beautiful and fascinating than cardinals singing in early spring or a goldfinch perched on a birdbath? Americans spend millions of dollars purchasing birdseed, feeders, and birdbaths to entice these creatures to our backyards. Of course, for the serious bird-watcher, binoculars are also a must to determine the specific color of the bird's eyes or the size of its beak or the color of its under wings.

One of my daily gratifications is to watch birds at my feeders. They often defend their territory by shrieking or diving at each other. In the spring I put out the hummingbird feeder weeks early just in case one of these little creatures decides to make an early debut. It is fascinating that such small birds can be so aggressive as they buzz away invaders near their feeder.

While the bee hummingbird is the smallest of all birds, weighing only 0.056 ounce, the ostrich is one of the largest. While Woodie and I were vacationing on the coast of Spain, one morning our host prepared scrambled eggs for breakfast, but she used only one egg to serve four people. It was an ostrich egg, which was larger than a grapefruit. An ostrich stands up to 9 feet tall and can weigh more than 350 pounds, so it produces gigantic eggs. Between the bee hummingbird and the ostrich come birds of every assortment, varying in size, color, and agility.

No one knows the exact number of birds, but it is estimated that 200 billion birds are on the planet at any given time. They are found everywhere there is land and even over the oceans, so why do we become alarmed regarding the demise of a few birds? In fact, sometimes birds such as pigeons become a nuisance. Talk to my neighbors about the sometimes pesky Canada goose that eats every green thing in yards and leaves its excrement in driveways. The Canada goose is increasing by 7 percent yearly and the wild turkey by 14 percent, so is there really a problem with the extinction of birds?

Though there is an increase in a few species of bird populations, as reported by the National Audubon Society, there has been a significant decrease in more than 20 species. The society owns a database for 550 species of birds and reports that 20 species have declined in population by half since 1967. Among these are the fence-sitting meadowlark, the rufous hummingbird, and the whippoorwill with its daunting call. In our country, there are 432 million fewer birds: a

A bird swoops over the North Sea.

decrease of 78 percent in the northern bobwhite and evening gros-
beak, a 73 percent decrease in the boreal chickadee, a 72 percent
decrease in the eastern meadowlark, a 68 percent decrease in the field
sparrow, and a 65 percent decrease in the grasshopper bird.[12] Other
significant decreases can be seen in the lark sparrow, the common
grackle, the horned lark, the little blue heron, and the ruffed grouse.
The northern bobwhite population dropped from 31 million in 1967
to approximately 5.5 million today. The common grackle dropped
from around 200 million to 73 million at present.

The causes of these decreases in the population of birds—including the extinction of some birds—are similar to the causes of the
decrease in other endangered species. By and large, human behavior
eliminates their habitats, such as forests and grasslands. Also the introduction of non-native species, over-hunting, and the use of
chemicals—such as DDT that killed so many birds in the 1960s—are
among the major causes.[13]

To assist us in being cautious regarding animals lest we contribute
to their extinction, we turn to Deuteronomy for the following verses:
"If you come on a bird's nest, in any tree or on the ground, with fledg-

A goat rests peacefully upon a sacred cow in the District of Uttar Pradesh, India.

lings or eggs, with the mother sitting on the fledglings or on the eggs, you shall not take the mother with the young. Let the mother go, taking only the young for yourself" (Deut 22:6-7).

Our Responsibility to the Fauna

Ezekiel reveals a message from God to the leaders of Israel, saying, "Ah, you shepherds of Israel who have been feeding yourselves! Should not shepherds feed the sheep? You eat the fat, you clothe yourselves with the wool, you slaughter the fatlings; but you do not feed the sheep" (Ezek 34:2-3).

God created order in the universe, including an array of diverse creatures, and we were commanded to exercise wise management of the fauna. However, often we have committed sins of omission and commission, both actions that have caused direct harm to animals and birds and actions that we have failed to take to help them. Sometimes the spiritually minded are not earthly grounded.

Animals are part of the circle of life. We find healing power in nature, and the fauna connect us to this force. We gain knowledge and derive pleasure from both domesticated animals and those in the wild.

More than 100 million people in the United States participate in wildlife-related recreational activities, so the commercial benefits from wild animals are enormous. The aesthetic benefits of observing animals are immeasurable, and with their biological diversity, they are considered an invaluable resource.[14]

We depend on the ecosystem to provide food, purify the air, and provide clean water, elements required by both humans and animals. We have come to the edge of creation and must not fall over the edge as a result of supporting our inclination for abundant living. If we destroy the earth, we could well destroy ourselves.

Questions for Reflection

1. What is meant by "The creation that is called to worship God is silenced"?
2. With one-third of living creatures in jeopardy, what is needed to protect them?
3. What are the primary reasons for the extinction of some birds?
4. We love our domesticated animals and spend millions each year on their care. How do we justify such expenditures when as many as two species of wild animals become extinct each day?
5. How are animals a part of the circle of life?
6. What is the difference between "endangered species" and "threatened species"?
7. Discuss some of the endangered species and what may happen if their habitats are destroyed.

Notes

[1] Johannes M. Boali and Casa Francescana Edizioni Porzuncola, *The Little Flowers of Saint Francis*, trans. W. Heywood (Assissi, Italy: Press la Tipolitografia Porziuncola S. Maria degli Angeli, 1982), 81–82.

[2] "Peace with God the Creator, Peace with All of Creation," message of His Holiness Pope John Paul II for the celebration of the World Day of Peace, 1 January 1990, http://conservation.catholic.org/ecologicalcrisis.htm (2 February 2005).

[3] Marcia Bungee, "Biblical Views of Nature: Foundations for an Environment Ethic," http://www.webofcreation.org/Articles/bunge.html (30 December 2008).

[4] Endangered Species Act of 1973, United States Fish and Wildlife Service (FWS), http://www.fws.gov/endangered/ESA35/ESA35DaleQA.html (30 December 2008).

[5] "Fact Sheet on Re-examination of Decision for 1998 Florida Black Bear Petition Finding," U.S. Fish and Wildlife Service, North Florida Field Office, http://www.fws.gov/northflorida/Florida-Black-Bear/reaxam-of-finding-fact-sheet-010804.htm (accessed 30 December 2008). See also "Florida Black Bear Removed from Endangered Species Candidate List," FWS press release, 8 December 1998, http://www.fws.gov/southeast/news/1998/r98-117.html (30 December 2008).

[6] Lauren Kurpis, "Why Save Endangered Species," http://www.endangeredspecie.com/Why_Save_.htm (30 December 2008).

[7] "Target Earth," Eden Conservancy, http://www.targetearth.org/eden_main.html (30 December 2008).

[8] Sharon Begley, Scott Johnson, and Julie Scelfo, "Cry of the Wild," *Newsweek* (6 August 2007): 20.

[9] Traci Watson, "Vanishing frogs tied to global warming," *Greenville* (SC) *News*, 12 January 2006.

[10] "NASA Study Finds World Warmth Edging Ancient Levels," Goddard Institute for Space Studies, 25 September 2006, http://www.giss.nasa.gov/research/news/20060925/ (30 December 2008).

[11] "Nature's Voice," Nature's Voice, Our Choice, newsletter, May/June 2007, 2–3.

[12] Seth Borenstein, "Some common American birds pops," *Daily Journal*, Seneca SC, 20 June 2007, 8A.

[13] Christine Crapanzano, "Bird declines have experts atwitter," *Greenville* (SC) *News*, 15 June 2007.

[14] Kurpis.

Chapter 5

THIS LAND OF OURS: AQUATIC LIFE AND CLEAN WATER

Circling over the Great Barrier Reef in a small seaplane was a memorable event as Woodie and I viewed one of the most majestic scenes within God's kingdom. Islands of turquoise and teal hues of coral were pinned just beneath the surface of the crystalline water. It was a feast for the soul. This natural wonder of creation stretches along the east coast of Queensland, Australia, for more than 1,200 miles, with some of the reefs reaching as far as 160 miles into the coastal waters. From the air we could see many of the almost 3,000 reefs that are a haven for marine life. We could also see some of the 600 islands that provide a natural habitat for wild birds and nesting turtles.

Our seaplane landed in the sloshing waves, and we boarded a pontoon to float over the coral reef. The pontoon had a window in the floor bottom where we could observe coral reefs and unique marine life up close. Coral is a tiny marine polyp of more than 400 variations that feeds mostly on plankton and grows in warm climates, providing a haven for thousands of sea creatures. Coral thrives in crystal clear saltwater and sunlight but does not flourish in polluted waters, which is one reason it is not found near seashores. The varying depth of the

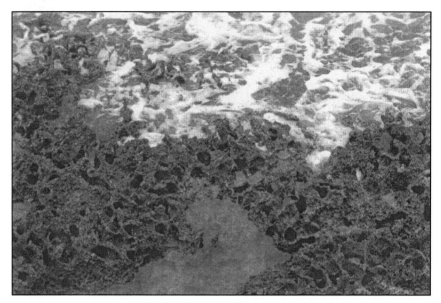

Dead coral reef on the coast of Cartagena, Colombia.

reef causes the color to be diffused, thus creating more hues than a field of flowers.

Swimming among the coral reef, I learned, were more than 1,500 species of fish, sea snakes, worms, and sea turtles. The fish were of the deepest hues of red, canary yellow, blue, and royal purple, some with stripes and others with dots and intricate lines on their bodies. They were of varying sizes and shapes, some gracefully beautiful and transparent, others grotesque with mouths as large as their bodies, and some with protruding eyes the size of eggs. The varying creatures of fish, snakes, and worms moved gently among the 500 species of ragged seaweed that swayed as schools of sea creatures gently flowed past them.

This Great Coral Reef is part of the sacredness of nature, inherited from the sea to be relished and protected. It is a revelation of creation. Hands could never create this terrestrial treasure developed through centuries as a majestic attribute to the kingdom of nature.[1]

As with other ecosystems, human activity poses a significant threat to this and other coral reefs. In fact, more than half of the coral reefs in the world are endangered. Fishermen in search of coral fish sometimes use ghastly techniques in trapping the fish, including dynamite, which devastates the reefs. There is also a considerable increase in the acidity of the ocean water as it absorbs carbon dioxide from the air. If countries would limit their fish export quotas, restrict or limit fishing in the coral areas, limit pollution, and regulate tourist swimmers and divers, they could lessen the destruction to coral reefs.[2] Scientists predict that by 2050, coral reefs could be rare.

Scriptural View of Living Water

"God said, 'Let the water under the sky be gathered together into one place, and let the dryer land appear.' And it was so. God called the dry land Earth, and the waters gathered together he called Seas. And God saw that it was good" (Gen 1:9-10). "And God said, 'Let the waters bring forth swarms of living creatures and let the birds fly above the earth across the dome of the sky.' So God created the great sea monsters and every living creature that moves, of every kind, with which the waters swarm, and every winged bird of every kind. And God saw that it was good" (Gen 1:20-21).

The psalmist also recognizes this creative power of God and offers praises for God's resourceful design:

> You set the earth on its foundations, so that it shall never be shaken. You cover it with the deep as with a garment; the waters stood above the mountains. At your rebuke they flee; at the sound of your thunder they take to flight. They rose up to the mountains, ran down to the valleys to the place that you appointed for them. You set a boundary that they may not pass, so that they might not again cover the earth. You make springs gush forth in the valleys; they flow between the hill, giving drink to every wild animal. (Ps 104:5-11)

Like the psalmist, Job recognizes God's majestic design, even as he anguishes over his own plight. God asks Job, "Who shut in the sea with doors when it burst out from the womb? Has the rain a father or who has begotten the drops of dew? From whose womb did the ice come forth, and who has given birth to the hoarfrost of heaven? The waters become hard like stone, and the face of the deep is frozen" (Job

38:8, 28-30). Also, "Who has the wisdom to number the clouds? Or who can tilt the waterskins of the heavens when the dust runs in a mass and the clods cling together?" (Job 38:37-38).

Jesus often illustrated parables using water as a symbol for the living water of God. He said to the disciples, "Follow me and I will make you fish for people" (Mark 1:17). A favorite story regarding Jesus and water is found in John as Jesus meets the Samaritan woman who came to a well to draw water. Jesus said to her, "Give me a drink." The Samaritan woman said to him, "How is it that you, a Jew, ask a drink of me, a woman of Samaria?" Jesus answered her, "If you knew the gift of God and who it is that is saying to you, 'Give me a drink,' you would have asked him, and he would have given you living water" (John 4:7-10).

Another favorite story in the early life of Jesus relates to the calming of a storm on the Sea of Galilee. He used this experience as an opportunity to teach the disciples about trust. Jesus climbed into a boat, with the disciples following him, and together they rowed out into the Sea of Galilee. Jesus, who was very tired, used these moments on the boat to sleep. Suddenly a storm appeared upon the horizon, and the waves began to rock the boat violently. The disciples became

Valencia Island, Ireland.

fearful and awoke Jesus saying, "Lord, save us! We are perishing!" Jesus replied, "Why are you afraid, you of little faith? Then he got up and rebuked the winds and the sea; and there was a dead calm" (Matt 8:23-26). The disciples were astounded at this turn of events; actually they were flabbergasted that this man could speak calmly to the winds and the waves, causing them to react peacefully. It was a moment of discovery, finding power and trust in Jesus.

Water is a powerful symbol in other Bible verses. John spoke to those in the desert of Judea saying he would baptize with water. John speaks similar words regarding baptism in Mark 1:8: "I have baptized you with water; but he will baptize you with the Holy Spirit." In Matthew 3:6 we find, "they were baptized by him in the river Jordan." Matthew 3:11 indicates that John said, "I baptize you with water for repentance, but one who is more powerful than I is coming after me." Matthew 3:16 reveals that John baptized Jesus: "Just as he came up from the water, suddenly the heavens were opened to him." John 4:1-2 states, "The Pharisees had heard that . . . Jesus is making and baptizing more disciples than John—although it was not Jesus himself but his disciples who baptized."

The Gospels of John and Luke both contain references to water symbolism. In John 13:5 Jesus pours water into the washbasin and begins to wash the disciples' feet. John speaks of "rivers of living water" in John 7:38, and in John 3:5 Jesus states that no one can enter heaven unless he is born of the water and the spirit. In Luke, Jesus says, "Do you see this woman? I entered your house; you gave me no water for my feet, but she has bathed my feet with her tears and dried them with her hair" (Luke 7:44).

In Corinthians, Paul addresses the believers regarding their state of infancy as Christians, saying, "I planted, Apollos watered, but God gave it growth" (1 Cor 3:6). Paul used this illustration of watering plants to reveal to Christians the concept that God's people are laborers in God's field.

Because creation was of God, God must dwell in the earth and in the waters. He gives individuals a special role to exercise care over his creation, which indicates that we should be not destroyers or polluters but rather preservers and protectors of our resources. To be in commu-

A woman pumps water in the rural area of the District of Uttar Pradesh, India.

nication with the environment and its valued resources, we must "till it and keep it" (Gen 2:15). God gave the standard, and we must maintain his directive. In disregarding creation care, not only do we destroy God's gifts, but we also diminish the food supply, especially for the poor, and remove sources of employment for the needy: "Truly I tell you, just as you did it to one of the least of these who are members of my family you did it to me" (Matt 25:40).

A Thirst for Clean Water

The Jordan River and the Dead Sea are biblical and ecological treasures for Christians. It is believed that Sodom and Gomorrah were located on the banks of the Dead Sea, that nearby Mount Nebo was where Moses allegedly first saw the promised land, and that the Jordan River was the site of Jesus' baptism. Today the Dead Sea and the Jordan River are in peril of drying up and have been polluted by human activity.

As we read the biblical description of the Jordan River, we find it a lush, refreshing 200-mile river in the desert area of Galilee. It was a river of clear, refreshing water, a means of transportation, and a source of water and fish. But today the river is quite different. Its water flow is less than 8 percent of its previous flow. It has been inundated with untreated sewage and other pollutants to the degree that the odor becomes stifling as one approaches. Contributing also to the plight of the river is the diversion of the river water in the early 1950s to Syria, Israel, and Jordan for agricultural and domestic purposes.[3]

As for the Dead Sea, the contaminated waters of the Jordan River now contribute to its dilemma. The fresh-flowing waters of the Jordan River were once the Dead Sea's source of replenishment. In biblical times the Dead Sea, which is the largest of the three seas that intersect the Jordan River, was known as the Salt Sea. With no exit flow and the seawater evaporating rapidly, there is a salt composition of 33 percent. The Dead Sea, 42 miles long, is the lowest point on earth at 400 feet below sea level and is shrinking by 3 feet every year, partly due to global warming and also from the lack of water flowing into it from the Jordan River. There is interest from Israel and Jordan in bringing a pipeline from the Red Sea, 125 miles east, to provide water for the Dead Sea, but there is no agreement as to how to pursue such a major project.[4] Not only has the water of the Dead Sea and the Jordan deteriorated in the last fifty years, but so has much of the quality of water in the world.

Though global warming is a contributor to the lack of clean water in the world, it is only one of several factors that play a role in the global water crisis. Others include wasteful irrigation (especially for people who lack sufficient infrastructure to utilize water properly for irrigation), domestic and industrial pollution, incompetence and corruption by politicians (especially in developing countries), and runoffs from farms and fields that use chemicals for their crops. So the real crisis may not be so much the lack of water as it is the need for proper water management.[5]

The lack of infrastructure for water was evident as I spent three weeks in 1996 in the developing country of Albania as a consultant to teachers of special needs children in an elementary school in Tirana. No drinking water was available in the school, but nearby was a water pipe that the children frequented at recess. A small stream of water ran

Students congregate at their school's water fountain. Tirana, Albania.

continuously from the pipe that was about 2 feet off the ground. A simple off-and-on faucet would have preserved thousands of gallons of water. The streets of Tirana were routinely wet and muddy due to leakage of the water lines, and one had to step cautiously at all times to avoid the mess.

The infrastructure was also poorly managed in the city of Allahabad, India, where I lived in 1994 in an affluent home of 25,000 square feet (where extended families with sons and their families occupied much of the house), a home with ten servants, yet where water was available only a few hours a day. My bathtub was filled by buckets of water. When I visited a beauty salon in the city, my hair was shampooed from a pan since water was too scarce to use from a faucet. In the slums of New Delhi, the lack of water is precarious. People wait for water trucks to arrive and sometimes pay ten times more for water than do the wealthy.

Though there is some infrastructure in most developing cities, I often observed in countries such as South Africa, Haiti, Peru, Kosovo, and Guatemala that wells were the primary source of water in rural areas, often located many miles from some residences. It is a common sight to observe women and children with hefty jugs of water atop

Rotarians help design a fish farm for poverty-stricken families of Lucknow, India.

their heads, walking long distances to secure water for their families. In India, I was told by a person in charge of rural education that more than half of the school-aged girls were unable to attend school due to their daily chores of securing water and gathering sticks for cooking.

Another major issue that hinders the quality of life for people in developing countries is open defecation, which is a routine practice, especially in slums and rural areas. Fifty percent of the world's people lack sanitation facilities, and when this factor is combined with the lack of clean water, diarrhea becomes the world's second-leading killer of children (after malnutrition). Estimates are that 1.8 billion children die each year due to inadequate sanitation and contaminated water. Though water may be available, it is often unsafe to drink. When clean water and sanitation are accessible, people are healthier, health costs are lower, people live longer, and life is enhanced. Safe water provides communities with the ability to enjoy food, own cattle, and grow crops, which in turn enhances the economy and quality of life.

Many of these unfortunate people have less than the minimum 5–13 gallons of water per day that is deemed necessary for living. Americans consume an average of 158 gallons per day with flush toi-

One-fourth of our beaches and lakes are unsuitable for
swimming and fishing. Garden City, South Carolina.

lets, showers and baths, dishwashers, automatic lawn sprinklers, and
our hefty meat diets (realizing that a thousand times more water is
required to grow beef than to grow grain).[6] Water, a necessity for survival, is scarce for one out of six people.

As we review the status of water in the world, we discover that 75
percent of the surface of the world is covered with water, but 97 percent of that water is brine. Two percent of the fresh water is
unavailable for consumption because it is in glaciers and ice or locked
in the ground, so that leaves only one percent of the world's fresh
water to supply the 6.6 billion people living on the earth.

As the world's population increases, so do the demands for additional water, especially for clean, safe water. In their book, *This
Moment on Earth,* John and Teresa Kerry reveal just how badly our
water resources have been fouled in the past few decades, especially the
rivers, lakes, and beaches in the United States. Kerry states that
100,000 of our nation's lakes and one-fourth of our beaches are mar-
ginal or unsuitable for swimming, fishing, or supporting marine
species. There were more than 18,000 closings and advisories for
beaches with unsafe levels of contamination in 2003, primarily related

to mercury. Also contributing to the problem was untreated human sewage and animal waste. Kerry further states that 846,000 miles of our rivers were under fish advisories in 2003 with recommendation for individuals, especially children and pregnant women, to refrain from eating the fish caught from these waters.[7]

Additionally, Kerry states that much of this contamination in streams and lakes occurred as a result of President Bush's decision in 2002 to reverse previous policies, permitting power plants to emit five times more mercury than formerly allowed. As a result, more than 100 power plants increased their mercury emissions into waters. In 2006 the EPA issued advisories for specific areas in forty-eight states, warning the public not to eat some varieties of fish from these waters.

Pollutants infiltrate our rivers and lakes as the rain carries some of the 80,000 chemicals used to rivers and lakes, along with gasoline and oil drippings from cars, household cleaning chemicals, water from chemicals and fertilizers in our yards, and rain runoff from sewage collection. Most water pollution can, however, be directly linked to agricultural practices when pesticides and manure are carried by rain into the rivers and lakes.

Hartwell Lake is usually eleven feet deep. Seneca, South Carolina.

Factories must share in the responsibility since over the years they have polluted and continue to pollute through releasing chemicals such as mercury and PCBs, which remain in the water for decades. The Mississippi River now has fifteen times more nitrates than it did previously as a result of agricultural runoffs, more than any other U.S. river. Polluting is a major concern, especially when it is estimated that 6 million Americans are sickened annually from drinking contaminated water.[8]

The cornerstone of water quality in the United States was provided by the Clean Water Act of 1972, which addressed primarily surface water quality and reduction of pollutants discharged into water and also assisted in the financing of municipal wastewater treatment facilities. However, since that time a great deal of attention has been directed toward discharges from such sources as sewage plants and industrial facilities. In the 1980s there was increased concern regarding polluted runoff, and today a more holistic watershed-based program includes the maintaining of water quality and restoring of impaired waters. This act, plus the involvement of states, is addressing water quality in hopes of resolving some of the critical water needs of our nation.[9]

Comfort comes in being connected with nature.
Winnie and Woodie, Newfoundland, Canada.

As I reflect on my childhood, I realize that I never was concerned that clean water would not be available, then or at any time in the future. My family, as well as our neighbors, had either a pump or a well with a bucket in our backyard. The water was sparkling, refreshing, uncontaminated, and always available. I recall the pristine nature of the creeks and ponds where I set "trot lines" or fished with a cane pole. I wandered over the woods, gathering hickory nuts and picking berries and persimmons while inhaling the pure fresh air. I did not have to worry about chemicals or sewage seeping into the streams or pollution from runoffs from fields and barnyards or whether the fish were safe to eat. The air and water was clean and healthy, especially in the country. It was a safer world to grow up in, and I was blessed.

Insulating Aquatic Creatures

In 2006 Woodie and I, along with two friends, traveled to Newfoundland and Labrador, located on the eastern shores of Canada, to view sea life, discover Viking history, and observe the countryside. Together these provinces have a population of about one-half million people. We were astonished to discover a fairly desolate province devoid of normal activity, and we soon learned that the lack of activity was described in one word—over-fishing. It is mind-boggling that the loss of codfish could cause the collapse of an entire economy, but that is exactly what has happened. The numerous villages and communities are practically ghost towns.

The cod, a sleek green and white fish, was the pride and lifeblood of the area until about ten years ago. The decline of codfish was gradual, but by 2003 Canada officials placed a moratorium on cod fishing due to over-harvesting. Over the past few years, people have begun leaving their small wood-frame houses and their childhood communities to seek employment elsewhere. They sought to sell their fishing boats, but the demand was almost nonexistent. It is a peaceful area, but one without much hope for restored employment for the fishermen in the near future. The number of children in one community school is a measure of how deserted the area has become: a decade ago there were eighty children in the school, but today there are only eight.[10]

One reason for over-fishing in Newfoundland and Labrador and other areas of the world is the enormous worldwide demand for fish, along with the evolution of super-technology in the fishing industry. One popular method of fishing requires the stretching of several heavy fish lines with thousands of baited hooks that sink to the bottom of the ocean. These lines may stretch a distance of several miles and catch more than the desired species. Other modern methods are echo sounders, helicopters locating schools of fish, and major netting devices. The government is presently permitting selected fishing in Newfoundland and Labrador with a specific quota and a limitation on the number of hooks used, making fishing almost an unprofitable occupation. When the limit for the week can be caught in a few hours, the few remaining fishermen must seek other sources of income. The codfish, the fish industry, and for the most part, the groups who have called the area home (some for more than 500 years) are almost gone.

Newfoundland is not alone in suffering the consequences of over-fishing. Many other countries are experiencing crises of equal proportions as their fish stock becomes almost nonexistent. Nearly a third of the world's fish are over-fished, according to a United Nations report in 2004, with the Atlantic being the hardest hit.[11] Should global fishing trends continue at the present rate, the worldwide fish stocks risk collapse as soon as fifty years from now.[12] Encouraging are the reports that certain areas in the Atlantic and Pacific that were almost depleted due to over-fishing are making a comeback thanks to government intervention.

Another consequence of some of the new technologies used in fishing is the endangerment of sea creatures caught unintentionally, such as sharks, sea turtles, unwanted fish that are dumped as trash, and sea birds that become ensnared in nets and fish lines. In 1995, New Zealand addressed the ecological crisis in its ocean waters and began marine reserves as an attempt to conquer over-fishing in its coastal areas. Today, thirty-one of these reserves stipulate no fishing or trapping, and prohibit anything to be taken from the waters. Their goals are "to educate the public, expand scientific knowledge and primarily protect and restore the ecosystem plus rebuilding biodiversity."[13]

Over-fishing is a real concern, but so are pollutants in the water. I spent a month in Puerto Rico several years ago, and one night I was taken for a boat ride to a bay near La Paguera on the Caribbean Sea. As I looked into the dark ocean waters, I observed fish glowing as if they were iridescent. I learned that the fish glow due to a chemical reaction between two microorganisms in the bay. These phosphorescent microorganisms attach to the fish when the water is stirred and cause the fish to sparkle as they swim in the dark waters. It was amazing.

Considering the conditions of the water and the consequences of over-fishing and other concerns such as global warming, is it any wonder that many species are included on the endangered list of North America? Among those on the list are reptiles and amphibians, including species of alligators, crocodiles, salamanders, and sea turtles. Among the large mammals are the dolphins, humpback whales, seals, and walruses.[14]

One of these unique aquatic creatures on the endangered list is the sea lion that exists on the western shores of the United States. Though I have seen sea lions in California, my closest encounter with them was on the Galápagos Islands. Early one morning we departed from our ship and boarded a panga (a small boat) to travel the short distance to Darwin Island (named after the naturalist Charles Darwin), one of the eighteen islands that make up the Archipelago of the Galápagos Islands. We encountered high waves that splashed and rocked the panga, but we were too excited to be alarmed.

As we arrived at the island, we saw dozens of sea lions sunning on lava rocks along the shores of the bay. Though the day was already sweltering, we were informed that we would climb to the top of these cliffs. We docked at the base of the cliffs, which were composed of serrated and uneven slippery volcanic rock perhaps 30 feet high. Climbing almost straight up, we used our hands and feet to claw from one rock to the next. However, it was worth the effort as we hiked for two hours over this pristine island, observing rare and fascinating birds that we had no idea even existed. On returning to our panga, I was well aware of the treacherous descent down the cliff. I slid more than I walked, only to encounter a sunning sea lion in the pathway.

The panga was moored to lava rock near the water's edge, but this huge sea lion, probably weighing 200 pounds, blocked my path. I was the first down the cliff, so I paused to get a closer look at this smooth, glossy creature. I observed that one of her flipper nails was missing, so I hovered closely over the sea lion that I thought was sleeping. I must have frightened her, and she flipped over rapidly toward me, growling and barking ferociously. Startled, I fell backwards, but the guide and Woodie prevented me from falling into the rocks. In spite of my zealous attention, the sea lion still would not remove herself from her sunny spot. We could not board the panga until she moved, clearing the pathway. Neither the guides nor tourists are allowed to touch sea lions or any of the animals to distract them. Finally, the sea lion rolled over toward the sea and dove gracefully into the water. Later, in a calmer atmosphere on a sunny beach, I was able to sunbathe with other sea lions. I can still hear in my mind the sea lions singing their poignant calls to their mates as the fiery sun warmed the island and cool breezes fanned the air.

Being among the sea lions, who are calm and friendly most of the time, I felt the freedom and lure of nature. The islands are painfully striking when the hands of humans have not marred the flora of the earth—with not even a tree destroyed. The volcanic rocks, speckled

A sea lion strikes a pose on the Galápagos Islands.

with blue-footed boobies and great frigate birds with red pouches, emitted a strong attraction for one ensnared by nature.

Though these sea lions of the Galápagos Islands live in a protected environment, at least 90 percent of large predator fish have been decimated, according to the Ocean Conservancy, an organization that has been active in ocean conservation since 1972.[15] The conservancy states that "the endless bounty of the ocean has come to an end. We've transformed once wild, untamed oceans into fish factories, garbage dumps, sewers, oil fields and express lanes for supertankers."

The conservancy also expresses alarm regarding the beluga whales, which are considered to be in danger of extinction. While traveling along the coast of Alaska, I observed these unique creamy-white, graceful whales as they romped and played along the coast of Cook Inlet. They have such a remarkable vocal repertoire that they have been referred to as "sea canaries" by sailors who have heard their various sounds of whistling, shrieking, and chirping. The beluga whale is smaller than most whales, averaging about 14 feet in length and weighing about 3,000 pounds. They have small melon-shaped heads, rather small tails, flippers, and a 6-inch layer of blubber that permits them to live in icy cold waters. Their population has fallen by 60 percent in the last 15 years due to over-hunting, industrial pollution, and the invasion of their habitat. The Ocean Conservancy projects that these whales will soon become extinct unless they are placed on the Endangered Species List by the National Marine Fisheries Service.

Moving closer to home, if global warming continues, a fish that could become nonexistent in South Carolina is the trout in upstate streams. Though streams are barely cool enough at present for trout to survive, if the temperature increases as much as two degrees, the trout will be unable to thrive, according to Tom McInnis, chairman of the conservation group Trout Unlimited.[16]

Another threat to fish is the invasion of foreign species into our waters. One example is the silver carp, according to the U.S. Fish and Wildlife Service. The silver carp, a native of Asia, was introduced into the United States in the early 1970s as a means of controlling algae in sewage lagoons and ponds, and these fish have now appeared in lakes and even in the Mississippi River basin. They grow up to 3 feet long, weighing as much as 60 pounds, and are "leapers." They leap into moving boats, injuring people and damaging equipment. Recently I viewed a newscast of a reporter attempting to photograph the silver carp for an article. Several carp leapt at least 6 feet out of the water, injuring people from several angles. This species competes with native species for food and habitat and threatens the fishing industry in certain areas.

Fishing for Answers

Though we read and hear of threats to aquatic life such as the invasive silver carp and the possible extinction of beluga whales, we must continually seek solutions regarding the recovery and safety of sea creatures. Looking for answers in the Bible is a good beginning.

Isaiah describes the joy of those who will be redeemed, prophesying, "For waters shall break forth in the wilderness and streams in the desert; the burning sand shall become a pool, and the thirsty ground springs of water" (Isa 35:6-7). But what happens to our world in the meantime? Could the thirsty ground burst forth with bubbling springs of water? Maybe not, but clean water could be made available to many

China's need for water increases with its population and industrial growth.
Nanchang, China.

more people of the world. We have the resources and the know-how to reverse or at least alleviate the water crisis and provide sustainable sea life, but we lack the will. Working together as individuals and cooperating as countries would make it possible to protect fish in the seas, lakes, ponds, rivers, and streams.

Water is the lifeblood of the world. Without it, neither we nor the animals or fish could exist, and if we continue on our current path, our children may one day find the ecosystem in shambles. On the other hand, if we collectively aspire to a cleaner, better bio-network, we can reverse some of the damage humans have caused.

I realized the significance of working together one evening as Woodie and I sat in a restaurant in Waikiki Beach, Hawaii. As we enjoyed a hassle-free evening with a sumptuous meal, we could not keep from gazing at the three-story saltwater aquarium in the adjoining lobby. The aquarium was 56 feet long and 35 feet wide and contained more than 70 varieties of resplendent rainbow-tinted fish. Some of the fish were smaller than my hand, while others spanned 4 feet in width, as did the stingray, and some 4 feet in length. The most amazing thing was that these many varieties of fish lived in such a manner that they did not invade each other's space, nor did they consume one another. I was told that because they were well fed, they had no need to devour each other.

As we observed this mass of diverse fish—green, yellow, red, slippery, scaly, grand, and dwarf—I saw it as a symbol of people living together in a concentrated environment, yet apparently at peace and respecting one another's space. We are more diverse than the seventy varieties of fish due to race, gender, nationality, and culture, and it is important that this multiplicity of heritage be preserved. However, in our diversity we still need love, support, and respect, which can be gained by listening to and caring for one another's concerns.

One way to address the needed repairs to the ecosystem is to influence the government in passing effective environmental policy. What we do in the United States affects the world. Though there is great diversity among people of the world, we all have similar goals. We all search for a healthy life, a clean environment, and abundant living. We need to rethink, adapt, seek solutions from one another, and become innovative in solving this complex puzzle of sustainability. What we do now is the future for the environment.

Questions for Reflection

1. Discuss some of the occasions when the word " water" is used in Scriptures.
2. How is water the "lifeblood of the world"?
3. With 97 percent of the earth covered with water and with only 1 percent of the water being potable, what can we do to keep it clean and available for the 6.6 billion people on earth?
4. Almost 2 billion children die each year due to inadequate sanitation and lack of clean water. Five to 13 gallons of water per day is considered necessary to sustain life; Americans consume an average of 158 gallons per day. Consider an experiment in which you use only 10 gallons of water in a day. What would this day be like? How would you do it?
5. Drought, pollution, and rechanneling water are drying the Jordan River and the Dead Sea. Can this happen to water bodies in America? Is it already happening?
6. What impact do over-fishing and damaging the coral reefs have upon the ecosystem?
7. Mahatma Gandhi said, "You must be the change you wish to see in the world." What changes will you make?

Notes

[1] Winnie Williams, *The Price and Privilege of Being a Woman* (Nevada City CA: Blue Dolphin Publishing, 2006), 129–30.

[2] John Kerry and Teresa Heinz Kerry, *This Moment on Earth* (New York: Public Affairs, 2007), 129.

[3] "Part of Jordan Tainted by Pollution," *Baptists Today* 24/1 (January 2006): 28.

[4] "The Dead Sea Is Dying," 28 November 2006, CBSnews.com, http://www.cbsnews.com/stories/2006/11/28/eveningnews/main2213458.shtml (30 Deember 2008).

[5] Bret Schulte, "A World of Thirst," *U.S. News and World Report*, 4 June 2007, 51–52.

[6] Ibid., 51–52.

[7] Ibid., 54.

[8] Ibid., 11.

[9] "Introduction to the Clean Water Act," Watershed Academy Web, http://www.epa.gov/watertrain/cwa/ (30 December 2008).

[10] Chris Carroll, "End of the Line," *National Geographic* 211/4 (April 2007): 92.

[11] "Still Waters, The Global Fish Crisis," *National Geographic* 211/4 (April 2007):49.

[12] Ibid., 112–13.

[13] Kennedy Warme, "Blue Oases," *National Geographic* 211/4 (April 2007): 79–81.

[14] Jane H. Imbre, "Endangered Animals of North America," http://www.kn.pacbell.com/wired/fil/pages/listendangerja.html (30 December 2008).

[15] "Beluga Whales," Ocean Conservancy, http://www.oceanconservancy.org/site/PageServer?pagename=fw_belugawhales (30 December 2008).

[16] Paul Alongi, "Warming trend could harm state's wildlife, experts say," *Greenville* (SC) *News*, 22 January 2006.

Chapter 6

THIS LAND OF OURS: THE ENERGY CRISIS

I am addicted to energy. I became more aware of this fact a number of years ago as I spent five weeks in Africa. For part of the trip I lived in a small rural village in Swaziland. At that time, I was, in general, oblivious to the primitive conditions that millions of the world's people confront daily. Most of the homes of the aboriginals in Swaziland were one-room round huts with dirt floors and thatched roofs, although some of the more affluent lived in small flat-roofed houses built of mortar and stone, usually without plumbing and electricity.

Two colleagues of mine and I had come to this remote village to teach Bible courses for adult students who wanted to become Christian workers among their people. I lived with a wonderfully kind African woman in a house that she had built herself, which consisted of a bedroom, sitting room, and small kitchen. There was a tiny bathroom, but we had to heat water in the kitchen to take a bath. The house had no electricity, so we used candles at night, since the wick for the lamp didn't work. Even though it was July, the weather was cold in this desert region that has opposite seasons from America. The sun warmed our house and the schoolrooms by day, but the nights were chilly without heat. As I slept on a cot in the room with my hostess, I vividly recall, even now, the restless nights I experienced when the wind whipped through the cracks in the windows, causing the curtains to dance. I would get up, put on my robe, and cover myself with a coat and sweater and pray that the roosters would soon crow to announce the arrival of morning. It was easy to get up at 5:30, since we usually retired soon after dark, at about 6:00 pm.

At 6:30 each morning, we went to the nearby common dining area to prepare breakfast for the students and their families, who lived on the premises. We fed twenty-five to thirty people for each meal. For breakfast we usually had mush made of brown and white maize, jelly sandwiches and bananas, and sometimes a milo beverage or tea—not very appetizing at an early hour, but it was filling.

Another trying experience of living in Swaziland was the lack of water. Wells are far from most families' homes, but where I lived there was insufficient wind for pumping water. Also, the process of cleaning the floor was new for me. The lady got on her hands and knees and used a small brush to remove dirt from the wood floor. I wondered why she did not use sagebrush, as many in developing countries do, but I was hesitant to ask. Although her house was clean, most of the huts had no windows and were dark and emitted pungent odors. The inconvenience that frightened me, though, was not having a clothes dryer. After I washed my clothes by hand, I hung them on a wire clothesline to dry. Now, that was not too difficult, except I was cautioned always to look in the tall grass under the clothesline for cobras lurking in the warm sun. I longed for my washing machine and clothes dryer at home.

One night in neighboring South Africa, I slept alone in one of these little round huts in Krueger National Park, where countless wild animals roamed. I appreciated that the hut had wooden floors, since most huts had dirt floors, and a single light bulb hung from the ceiling. When darkness arrived, I heard the elephants braying, the lions roaring, and all kinds of real and imaginary animals of the jungle prowling. I knew they might walk right into my room, especially since I could not lock the door. I looked and listened most of the night for snakes in the thatched roof over my bed and, again, prayed for daylight. The sunrise was the most beautiful and blessed sight ever.

After these experiences, I knew for certain that I was addicted to energy, that is, everything that is operated by energy—electric lights, warm houses, food cooked on a gas stove, telephones, washers and dryers, and even the television. I wondered if I were capable of making adjustments in my lifestyle to conserve energy, to help reduce pollution so that those in Africa might have a better life.

Usually the first step to correcting an addiction is to admit that you have it. Dell Isham, director of the South Carolina Chapter of the Sierra Club, commented that we often say, "I know that I am addicted, just give me another fix while I think what to do about it."[1]

Does justice prevail when some nations are affluent and yet others have so little? It is impossible to predict with certainty how climate change will affect some countries, but it will not be uniform for each country. Certainly the poor, who have contributed the least to global warming, will be affected the most by it. People living in low-lying coastal areas in substandard housing will experience the thrust of dangerous hurricanes and flooding more extensively than those living in more substantial housing in developed countries. People in the Sahara Desert of Africa will most likely suffer more than others from lack of water, higher temperatures, high winds, and erosion. My prayer is that I will see the injustices of the poor not only with my eyes but also with my heart.

In Job we find concern for justice as Job's friends talk of God's wondrous works and emphatically state that the Almighty in his justice and great righteousness does not oppress. Job's friends ask him to pause for a moment, listen, and consider the wonderful things that God does:

Hear this, O Job; stop and consider the wondrous works of God. Do you know how God lays his command upon them, and causes the lightning of his cloud to shine? Now, no one can look on the light when it is bright in the skies, when the wind has passed and cleared them. Out of the north comes golden splendor; around God is awesome majesty. (Job 37:14-15, 21-22)

My healing garden. Seneca, South Carolina.

We cannot love our neighbors as ourselves if we do not take into account environmental concerns. We put love for our neighbor and care for creation into action by respecting the flora and fauna. Paul writes, "All things are lawful, but not all things are beneficial. All things are lawful but not all things build up. Do not seek your own advantage but that of others" (1 Cor 10:23-24). Developing an environment ethic of preserving and caring for this world for everyone, including the poor, our children and grandchildren, and ourselves, should be a worthy goal for Christians.

Depleting our Resources

As I walked down streets in India, I observed women following cows as these sacred animals meandered aimlessly through the villages. The women followed the cows until the cows dropped dung. The women would stop and rub sandy dirt on their hands as one would use flour before touching something sticky. Then they would scoop up a handful of the dung and form patties. When they had several of these patties in their baskets, they took them home and stuck them on an outside wall to dry. Later, they used the dried dung as fuel to cook a meal for their families.

I watched in Peru and Africa as women and children walked long distances to gather sticks for cooking fuel and then returned to their homes with heavy loads on their backs. I watched Chinese people as they burned coal in small pots in their yards, using the fire for cooking and heating water. These are some of the 2 billion people who cook meals with firewood, dried dung, or coal. One and a half million of these people will die young from breathing in the fumes of the various fuels. Most of the world uses fossil fuels such as gas, oil, and electricity generated largely by coal. China builds on average one coal-fired power station a week.

Though the United States is the largest polluter in the world, China, with its vigorous economic development, will probably soon surpass the U.S. as the major polluter. However, with the United States outsourcing much of its manufacturing and then receiving large numbers of imported items from China, we are a major indirect contributor to China's emissions of greenhouse gases. Though the United States is considered one of the wealthiest and most powerful nations in the world, we are one of the countries that lags regarding renewable-energy production, according to James Speth, dean of the Yale School of Forestry and Environmental Studies.[2] In comparing the environmental health of 133 nations, the United States ranked 28th in the world, following such countries as New Zealand, Austria, and the Czech Republic. Several criteria were included, such as air, water, biodiversity, and energy. When clustered with similar countries, the United States landed close to the bottom, at number 12.

As a result of smoke stacks and tailpipes belching fossil gases into our atmosphere for a century or more, we have played a part in causing the planet's thermostat to rise to dangerous levels. The major greenhouse gases of coal and oil have played a role in threatening our ozone levels. Other less harmful sources of energy include natural gas and nuclear energy.

Sources of Energy

Typically, oceans or plants absorb half the contaminants from fossil fuels that are emitted, but the rest remain stored in the atmosphere. Briefly, the major sources of energy and their impact on the world of energy are reviewed below.

Coal

Though coal-fired generators, developed in the 1890s, brought us electricity with all its amenities and changed the American way of life, the coal-powered industries have now become so numerous that coal and oil are the leading industrial sources of air pollution. Though coal is the cheapest of the major fuels used in the United States, it is the worst polluter, even more so than oil. According to Al Gore, coal "puts mercury into our air and water, soot into our lungs, destroys our mountain tops and kills our workers." Gore also states that coal and oil together, with their high carbon content, produce 70 million tons of carbon dioxide every day.[3]

When sulfur dioxide and nitrogen dioxide are released from coal industries, they mix with oxygen and water in the air to form acid rain, which may travel long distances by the wind. Also, some of the sulfur dioxide released during the production of electricity at power plants creates small particles that create haze.[4] Sulfur dioxide emissions are more prevalent in areas where paper mills and wood pulp are processed, in areas close to power plants, and in high-traffic areas.

Oil

Although oil provides many of the comforts for our lives, especially gas for vehicles, it produces almost one-third of the carbon dioxide linked to global warming. It also provides for the operation of many industries on which we depend, but the greenhouse emissions from oil are detrimental to our health, our wildlife, and our environment, including drinking water.

According to the National Resources Defense Council,[5] there is more spillage of oil than any other chemical. The oil industry reported 4,534 toxic spills across Alaska's North Slope and the Bedford Sea from 1996 to 2004, which averages more than one spill a day. Whether pipelines, ships, or trucks transport oil, all pose threats of spillage. The devastating consequences of spillage include contamination to groundwater and serious harm to wildlife. Also, the dumping of oil, sometimes deliberately, in drainage systems is perilous. Even drilling in the ocean has consequences for the environment, such as the dumping of solids and gases into the ocean from oil rigs, dumping machinery, domestic sewage, drill cuttings, and drilling fluids. Another problem involving oil is the disturbance of seabeds when rigs are placed in the ocean, causing fish that are exposed to oil to develop tainted skins.

Most people can recall the Exxon Valdez oil spill in 1989 that caused 11 million gallons of oil to seep into the water, spreading along 470 miles of the Alaskan coast. This one spill devastated the ocean and seashores for years, in spite of efforts to clean up the oil. It caused the death of fish and other wildlife, including many species of birds.

The Natural Resources Defense Council also states that noise from oil and gas exploration has caused whale stranding and hearing loss in marine mammals that depend on sound to navigate, hampered fish in finding food, and hindered mates in locating each other. The NRDC further insists that oil exploration drives away or even kills the fish that marine mammals depend on for survival. The oil spills in the Beaufort Sea along the Alaskan coastline pose an extraordinary problem. In these icy waters, the oil could place an irremovable layer of thick toxic oil on glaciers for generations to come, suffocating whales and porpoises and decimating thousands of migratory and native birds (who ingest the oil while preening), and also displacing and killing polar bears and seals. Oil exploration threatens wildlife when roads and drilling rigs are constructed, when equipment, materials, and workers are transported into the forest habitat.

Even though there are people who ridicule those who embrace creation care, there *is* a crisis regarding greenhouse gases. Some involved with the oil and coal industries have promoted the idea that global warming is a hoax. Simply put, they stand to lose money if alternative sources of fuel are developed. For example, ExxonMobil announced early in 2007 that it had the largest annual profit for the preceding year, 2006, of any corporation in the history of the United States. According to Al Gore, ExxonMobil has funneled nearly $16 million between 1998 and 2005 to a network of 43 advocacy organizations that seek to provide information to confuse the public on global warming science.[6]

While some jeer at our concerns, we would do well to read what Simon Peter said to the Christians concerning false prophets: "First of all, you must understand this, that in the last days scoffers will come, scoffing and indulging in their own lust . . . saying all things continue as they were from the beginning of creation" (2 Pet 3:4-5).

Nuclear Energy

Nuclear power, which provides about 20 percent of our energy, produces no carbon dioxide, the main greenhouse gas. It is one of the cleanest of all energy sources but does have negative effects upon rivers, lakes, and reservoirs. Nuclear fission requires huge amounts of cool water to keep reactors operating at safe temperatures; the reactors then return hot water to lakes and streams. The hot water does cool relatively quickly, but fish and other animals often are killed when they become trapped against the screens covering the cool water intake pipes. Further, as climate warms the cool waters, nuclear power plants are less able to deliver their energy and often have to curtail output or shut down the reactors. Inland sources of water grow scarce, most seashores do not welcome nuclear stations, and power is becoming inefficient as it is transported long distances. With two-thirds of nuclear plants now positioned on lakes and rivers, there is a real concern for finding sufficient cool water for nuclear energy. Another disadvantage for nuclear energy is that it requires large subsidies to sustain its operation.[7]

Natural Gas

Natural gas is the cleanest of the fossil fuels, thus having a lesser impact on the environment than the others. Natural gas, which is domestically abundant, is composed primarily of methane when burned. It has fewer carbons, nitrogen oxides, and sulfur dioxides, and lower particulate ratios than do coal and oil. The combustion of natural gas releases very small amounts of harmful pollutants, as compared with oil and coal.[8] However, small amounts of natural gas are lost in production and during distribution. Losses may be less than 2 percent in modern, well-maintained systems but slightly higher in old and poorly maintained systems.[9]

Some naysayers state that the economy will be wrecked if we move toward renewable sources of energy. On the other hand, reducing dependency on foreign oil and developing a clean energy strategy would spur investment in ethanol and biodiesel plants, wind turbines, hydrogen fuel cells, and new technology, according to Minnesota governor Tim Pawlenty, 2007 chairman of the National Governors' Association. In his first act as chairman of the NGA, Pawlenty pointed out that "our country is too dependent on imported sources of energy,"[10] and 90 percent of the 2007 governors in their state of address speeches highlighted policies on the environment.

In addressing environmental issues, Bill Moyer said in an interview with *Grist Magazine*, "The environment will kill us. The environment is a personal issue, an emotional issue and one of our most cherished values. It is a deeply moral issue, a moral imperative, that we need to save what is not our own so that those who come after us can have a life. It stuns me that the people in power can't see that the source of our wealth is the earth."[11]

A Vision for Energy

With the anticipation of 9 billion people living in the world by 2040, there will have to be some innovative means of providing new sources of energy, especially if we are to heal the ecological wounds that surround us. Saying that we are going to fix our energy sources in the future is not sufficient. They need fixing now. It took us a long time to arrive to get where we are, and no matter what we do, the environment will not improve rapidly, even with our best efforts. However, by doing nothing, we imperil our health and our natural resources and risk the extinction of much of our flora and fauna.

There has been innovative work toward the goal of reducing fossil fuels and making our country energy independent. Here are a few of the developments in seeking sustainable fuel replacement:

Solar Energy

Scientists at the California Institute of Technology are boosting the efficiency of solar panels. Solar power is one of the best alternative energy sources; it is abundant and could provide cheap energy to supplant fossil fuels.[12]

Wind Farms

Wind power is a good source of renewal energy that will aid in the reduction of air pollutants. The downside of wind farms on land is the noise factor and the fact that clearing and soil disruption (which may in turn cause erosion) are necessary to install wind turbines. Wind farms could generate up to 7 percent of the needed electricity in the U.S. in fifteen years. At present, even though wind farms are in thirty-six states, they produce less than 1 percent of the national electric

Wind turbines harness energy at Rotterdam, Holland.

power, although there are 3,000 or so wind turbines northwest of Palm Springs, California.[13] The lack of wind to power the windmills in some areas is problematic, as are their spinning blades, which threaten birds and bats. Creatures most vulnerable to the blades are night-migrating songs birds, bats, and predatory birds such as hawks and eagles. In placing wind turbines, there is a need to avoid areas that are high risk for birds.

Oceans, Tides, and Wind Currents

The oceans, tides, and wind currents can create energy to generate electricity, according to Roger Bedard, Ocean Energy Leader for the Electric Power Research Institute. He states, "The technology for harnessing marine power appears to have the potential as a new source of energy." Spinning turbines are secured to bobbing buoys with cylinders to convert ocean or river movement into electricity. The process has won preliminary approval from the Federal Energy Regulatory Commission. There is enough interest that the commission has issued twenty preliminary permits, while thirty-five permits are pending for wave and tidal projects. Most of this interest comes from the West

Waterfall demonstrates nature's potential for providing energy.
Yellowstone National Park.

Coast, Florida, and New England.[14] Bedard also states that there are indications that in fifty years or so, perhaps 10 percent of U.S. energy could be provided by wave and tidal energy.

An encouraging fact is that presently about half of the states require utilities to mix traditional power with some alternative energy. There are indications that water power appears to have some distinct advantage over wind power in that it can produce forty times more power than windmills and is more predictable than wind. Oregon has high waves, reaching about eleven feet in the winter, which would provide enough energy to generate sufficient electricity per yard of wave-crest to power about thirty-eight homes.

Agriculture

The process for utilizing corn and similar feedstock for ethanol production is leaping forward. South Dakota has thirteen operating ethanol plants and five more under construction and hoped to produce one billion gallons of ethanol by mid-2008.[15] Another process in the experimental stages is turning cellulose to ethanol. It is a process of converting cornstalks, corncobs, or other agriculture byproducts into ethanol. Other possible materials for ethanol production are pine needles and underbrush.

The Earth's Center

The center of the earth is also being explored as a source of energy. Switzerland engineers recently drilled into the earth's crust to tap its inner heat and determine if it could be a new source of energy. The idea is to send water to the earth's center about three miles underground, then retrieve the super-heated water to generate power for making electricity. The energy would be clean, quiet, and virtually inexhaustible. The resource base for geothermal is enormous. The downside is that it would be expensive and could possibly cause earthquakes. Scientists at the Massachusetts Institute of Technology say that to drill a hole three miles deep for such a purpose in the United States could cost 7 million dollars.[16]

Storing Carbon Dioxide

In West Virginia, scientists are studying the idea of drilling more than 9,000 feet below a coal-fired power plant to see whether layers of rock can provide a repository for vast amounts of carbon dioxide that is released when coal burns.[17]

Hybrid Cars

Much attention has been focused on hybrid cars in the past couple of years. Ford Motor Company's CEO Bill Ford says that by 2010, half of Ford's models will be available as hybrids, and they expect to sell 250,000 cars that year. Toyota expects to sell 600,000 hybrids a year by that time. A hybrid locomotive is also on the horizon. In fact, one now being used to shunt rail cars to different tracks in the freight yard is made by a Canadian company that hopes to reduce greenhouse gas emissions by as much as 70 percent.

In 1931 Thomas Edison said, "I put my money on the sun and solar energy. What a source of power! I hope we don't have to wait until oil and coal run out before we tackle that."

Environmental Ponderings

Many American are "going green" on their own, depending on their lifestyles, their interests, their economic base, and their belief system. A Gallup Poll in March 2007 found that more Americans than ever, 60 percent (up from 48 percent a year prior), believe that global warming has begun to affect the climate. In addition, an even larger group believe that it will cause major or extreme changes in climate and weather during the next fifty years.[18] Though it is hard to admit, Christians are responsible for some of the harm to our universe, even though it may have been unintentional.

As consumerism has crept into our lives and economic gains have taken precedence over the welfare of our ecosystem, we have lost sight of some of the larger purposes of life. It appears that we fail to understand the moral implication underlying our ecological plight.

Though we have responsibility for our individual behavior, I am convinced that our government should establish and enforce appropriate regulations to protect us from environmental contamination—from harmful chemicals in our food, atmosphere, and water. Internationally, there have been attempts at setting standards for improving our earth, although the United States, presently, has not taken environmental hazards seriously. As citizens, we need to insist on more effective measures from our political leaders regarding a cleaner and safer environment.

For example, emissions rise as does the population, and in the Southeast gas emissions rise sharply due to rapid growth. There are more cars and coal-fired power generators, and yet our local and national governments have failed to acknowledge what is happening. The United States Public Research Group reported that the Southeast carbon dioxide emission jumped 31 percent over the period from 1990 to 2004, versus an 18 percent jump nationwide.

Providing a safe haven for Americans and also saving the biodiversity of our land is not an easy task. There is not one simple solution. Near my home, a small mountain has been removed. It took about eight or ten years to whittle it down, but now on the land sits a thriv-

A fragile habitat. Killarney, Ireland.

ing shopping center. Where once stood the green mountain and forest, there are nine stores, and a habitat for wild animals has been lost. Once you lose a mountain, it cannot be regrown like a forest. The answer to population growth and the need for shopping centers is complex. However, when a forest is destroyed, you lose the biodiversity of the animals, plants, and trees, and such damage is irreversible.

What can we do? Psalm 121:1 says, "I will lift up my eyes to the hills—from where will my help come? My help comes from the LORD, who made heaven and earth." King Solomon encourages us to appreciate the present and anticipate the future: "See the winter is past; the rains are over and gone. Flowers appear on the earth; the season of singing has come, the cooing of doves is heard in our land" (Song 2:11-13). We must have hope that individually and collectively, we can regain much of what God has given us and thus protect the future.

Pope John Paul II, addressing an environmental conference in 1982, said, "Respect for life, and above all, for the dignity of the human person, is the ultimate guiding norm for any sound economic, industrial, or scientific progress. No peaceful society can afford to neglect either respect for life or the fact that there is integrity to creation." He also said, "We live in a harmonious universe and we must respect it. Today, the dramatic threat of ecological breakdown is teaching us the extent to which greed and selfishness—both individual and collective—are contrary to the order of creation, an order which is characterized by mutual interdependence"[19]

A writer for the *SC Sierra* newsletter adds this challenge: "We can fix the problem through education, legislation, litigation, and can reset the direction of the human race. We need a conservation ethic. We can maintain comfortable lifestyles and that of future generations by recognizing our responsibility to protect the nation's resources that God has given us."[20] Though we may be able to maintain a comfortable lifestyle, we may need to consider fewer frills, more moderation, and discipline, as well as a spirit of sacrifice.

Thomas L. Friedman in *The Lexus and the Olive Tree* says, "In the next decade, if globalization continues to bring more and more people into this lifestyle, and if we cannot learn to do more things using less

Growth abounds following devastating wildfires. Glacier National Park.

stuff, we are going to burn up, heat up, pave up, junk up, franchise up and smoke up our pristine areas, forests, rivers and wetlands at a pace never seen before in human history."[21]

Among the global issues that need to be addressed is the issue of poverty. In order to survive, many poor people in developing countries find it necessary to sell and exploit their natural resources in order to survive.

We are connected to nature. Contentment and recuperative energy come as we bond to nature and seek healing between our spirit and the natural world. Aside from nature, where else can one find such enchantment and awe? God allows us to be enveloped in the wonder of mountains and waters and the beauty of wildlife. In our appreciation for these gifts, we become fully alive and honor God's provision of the miracle of nature. We must preserve nature with sustainability—that is, supply our present needs yet not compromise the needs of the future.

The wonderful Creator holds the world close. He allowed us to become a part of his world. He is the Shepherd who takes care of us on the high mountains and in the deep valleys. He gives us strength to

Winnie visits with a Guatemalan family.

face the hardship of each day. In turn, we are obligated to give our best in protecting that which he has given to us.

Questions for Reflections

1. What is the relationship between God's word, God's mercy, and honoring the creation?
2. What is an environmental ethic? What is a carbon footprint?
3. Discuss the negative effects of fossil fuels such as coal, oil, and gas. What are alternative solutions?
4. How does outsourcing of products to developing countries contribute to global warming?
5. What impact will China have on the world's environment as they build an average of three coal plants a week for electricity?
6. How do we supply our present needs and not compromise the future?
7. Why is the U.S. ranked twenty-eighth in the world regarding renewable-energy production?
8. What are the implications of Bill Moyer's statement, "The environment will kill us"?

Notes

[1] *Congaree Chronicle* (newsletter of the Sierra Club, South Carolina chapter) 29/2 (2007).

[2] Dashkr Slater, "We're No. 28," *Sierra Magazine* 91/3 (May/June 2006): 9.

[3] Al Gore, *The Assault on Reason* (New York: Penguin Press, 2007), 200–201.

[4] "What Is Acid Rain?" EPA Acid Rain, http://www.epa.gov/acidrain/what/index.html (30 December 2008).

[5] "Stop Shell from Drilling this Summer Right off the Coast of the Arctic National Wildlife Refuge!" newsletter of the Natural Resource Defense Council, New York, 2007.

[6] Gore, *Assault on Reason*, 200–201.

[7] James Kanter, "Global Warming Imperils its Potential Solution," *Herald Tribune-New York Times*, 21 May 2007.

[8] "Natural Gas and the Environment," NaturalGas.org, http://naturalgas.org/environment/naturalgas.asp (30 December 2008).

[9] Mark Delucchi, "Appendix E: Methane Emissions from Natural Gas Production, Oil Production, Coal Mining, and Other Sources: An Appendix to the Report, "A Lifecycle Emissions Model (LEM)," Institute of Transportation Studies, Davis CA, November 2003, p. 2.

[10] "Incoming NGA Chair Unveils New Initiative," press release, Nation Governors' Association, Traverse City MI, July 2007, http://www.nga.org/portal/site/nga/menuitem.6c9a8a9ebc6ae07eee28aca9501010a0/?vgnextoid=074d60ca8c1c3110VgnVCM1000001a01010aRCRD (30 December 2008).

[11] David Roberts, "Faith, Hope, and Clarity," Interview with Bill Moyer, *Grist*, http://www.grist.org/news/maindish/2006/10/05/moyers/ (30 December 2008).

[12] *AARP The Magazine*, July–August 2007, 54.

[13] Randolph E. Schmid, *Desert Sun*, 5 May 2007, www.thedesertsun.com.

[14] Paul Davidson, "Catch a Wave, Throw a Switch," *USA Today*, 19 April 2007.

[15] Dan Daily, "Facilities Turning Wood into Ethanol," Smallwood Utilization, http://smallwoodnews.com/Forum.

[16] "Companies Hoping to Exploit Earth Heat as Energy Source," *Greenville* (SC) *News*, 5 August 2007, 10A.

[17] Marin I. Hoffert, *AARP The Magazine*, 2007.

[18] Patrick O'Driscollo and Elizabeth Weise, "Doing the right thing isn't easy, even for those who want to," *USA Today*, 19 April 2007.

[19] "Peace with God the Creator, Peace with All of Creation," message of His Holiness Pope John Paul II for the celebration of the World Day of Peace, 1 January 1990, http://conservation.catholic.org/ecologicalcrisis.htm (2 February 2005).

[20] Dell Isham, "Can We Have a Word about Energy?" *SC Sierra* (March 2006).

[21] Thomas L. Friedman, *The Lexus and the Olive Tree: Understanding Globalization* (New York: Anchor Books, Random House, 2000).

Chapter 7

MAKING GREEN
FOOTPRINTS

In the sixth century BC, when the Babylonians conquered the Israelites in Judah, they enslaved the people. It was a dark and distressing time, but God sent a prophet to the people, saying,

> I am the LORD, who opens a way through the waters, making a path right through the sea. I called for the mighty army of Egypt with all its chariots and horses, to lay beneath the waves, dead, their lives snuffed out like candlewicks. But forget all that—it is nothing compared to what I'm going to do! For I'm going to do a brand new thing. See, I have already begun! Don't you see it? I will make a road through the wilderness of the world for my people to go home, and create rivers for them in the desert! The wild animals in the fields will thank me, the jackals and ostriches too, for giving them water in the wilderness, yes, springs in the desert, so that my people, my chosen ones, can be refreshed. I have made Israel for myself, and these, my people, will someday honor me before the world. (Isa 43:16-21, LB)

The prophet understood the hopelessness and despair of the troublesome days for the Israelites, but he wanted them to visualize the future. He said, "God will do a brand new thing—there will be roads

in the wilderness and he will create rivers for them in the desert."
Isaiah encouraged the people to stop looking backward and start looking forward to something new, to have foresight and dream of better
days to come. God's promises came to pass for the Israelites.

Just as God pointed the way for the Israelites, he will do so for us.
God is still a God of innovative and fresh happenings. By complying
with his guidance, we can help transform events of the world. We have
developed dreadful habits that injure God's creation. It is a tremendous challenge to alter our lifestyles, but we have a duty to do so. Our
environmental footprints will be part of our legacy.

Even though debate over preserving our ecosystem continues,
concern is not so much about *if* changes are happening in the environment, but *when* they will happen. What we need is the will to see that
positive change happens. Mahatma Gandhi said, "It's not too late at
all. You just do not yet know what you are capable of." And Jonas Salk
stated, "The reward for work well done is the opportunity to do more.
Our greatest responsibility is to be good ancestors."

Protecting the environment begins with all of us learning to love
and respect nature and loving our neighbors as we love ourselves. If we

The lotus flower is a symbol of purity and great beauty.
Brookwood Gardens, Murrells Inlet, South Carolina.

love those around us and know that what we do will affect them, we will be more cognizant of our actions. We must commit to a healthy environment as part of our belief in God as the Creator. After all, as a coalminer says in the movie *Silver Spring*, "We think we can wound the earth and walk away, but one day the bill will come due."

Ways to Promote Creation Care

Caring for creation means reducing the amount of "stuff" we consume, recycling and reusing that "stuff" we do consume, conserving and respecting our resources. What we eat, how we travel, how we shop, how we garden, and how we seek to make our homes energy efficient can make for a greener earth. Many of the ideas listed below are not original but have been gleaned from numerous sources such as newspapers, magazines, the Internet, and communication with others.

At Home

- Use nontoxic and biodegradable products for house cleaning such as Shaklee Get Clean H^2 products. (Order from shakleegetclean.com.)
- Purchase containers for recycling and *use* them.
- Locate the nearest recycling center. Recycle aluminum and metal cans, glass, plastic, newspapers, and any other items your recycling center will accept. (Producing paper, glass, and metal products from recycled materials creates 70–90 percent less pollution than creating the same products from raw material.)
- Make compost from leftover food or use it to feed the birds and wild animals. (Some cities require families to have compost containers.)
- Use cloth dishtowels and cloth napkins instead of paper.
- Recharge batteries and recycle cartridges.
- Use fewer paper products such as coffee cups and paper plates.
- Don't leave water running while you brush your teeth.
- Hang clothes in the sun to dry. (By not using a clothes dryer, the average family could save up to $85 per year in energy costs.)
- Take shorter showers.
- Use plants indoors to help clean the air.

- Do not flush medications down the toilet (especially estrogen, which can cause deformities in fish).
- Have your utility company provide a home energy audit to determine how to improve efficiency.
- Caulk windows to plug air leaks for more energy efficiency.
- Select energy-efficient home appliances.
- Clean or replace furnace and air conditioner filters as recommended.
- Replace incandescent light bulbs with compact fluorescent bulbs. (Using these bulbs will reduce electrical cost by 75 percent. While initially more expensive, these bulbs last up to ten times longer than incandescent bulbs. If every family in America replaced just one light bulb, it would save more than $600 million in annual energy costs.)
- Turn the hot water thermostat down. (Each degree you lower your water temperature in the winter can reduce energy bills up to 39 percent.)
- Wrap your water heater in insulation (if more than five years old).
- Insulate your home. (Insulating your home can reduce home energy consumption as much as 25 percent.)
- Cover windows with curtains to keep out summer heat.
- Turn down thermostats in the winter for less heat and raise air conditioner thermostats in summer for less cooling. (A programmable thermostat allows you to set optimum temperatures automatically to optimize conservation.)

In the Garden

- Cultivate a garden and connect to nature and the earth.
- Carefully manage fertilizers and insecticides.
- Use black and white newspapers (no colored print) for mulch and compost (see above advice on composting).
- Plant many trees in your yard and neighborhood.
- Make compost of leaves, grass, and other garden material.
- Prevent mud and fertilizer from seeping into streams and natural areas.

While Shopping

- Buy food in bulk. (Help eliminate excessive packaging by dividing bulk food items among reusable containers at home.)
- Use tote bags to shop, especially at the grocery store. Always leave a tote bag in your car for unexpected purchases. (I make my own grocery bags from colorful, sturdy materials.)
- Choose rechargeable products over disposable products, such as batteries.
- Buy fresh rather than frozen vegetables. (Frozen vegetables require ten times more energy to produce than fresh.)
- Shop at farmers' markets for fresh food. (You save the transportation cost from long-distance handling and produce is fresher. It also encourages farmers to sell locally and is usually cheaper.)
- Consider buying organic foods, which are most likely free of pesticides. Read labels.
- Look for less packaging for items.

At Church

- Begin a church project to accept recyclables such as newspapers and aluminum cans that can be sold as a fundraiser.
- Request your church kitchen workers to recycle cans, glass containers, plastics, and paper products.
- Eliminate paper products at meals. Instead, use dishes, silverware, glasses, and mugs.
- Ask office staff to recycle paper. Recycle Sunday school literature.
- Request that Sunday school class members bring coffee mugs rather than using disposable cups.
- Distribute leftover food to needy families.
- Request that church custodians use nontoxic and biodegradable products for cleaning.

In the Community

- Write congressional and state leaders requesting that attention be directed toward designing green government buildings, developing in-state biofuel production, requesting clean air standards, and perhaps offering tax credits for more efficient vehicles.

- Support legislative efforts to safeguard the environment, such as limiting oil drilling, coal mining, and clear cutting of forests.
- Write letters to local newspapers and magazines regarding your environmental concerns.
- Talk with your city planning commission regarding zoning for subdivisions to lessen the impact on animal habitats and the environment.
- Find out how your community planners adhere to environmental standards for commercial development. How often are exemptions to these standards granted?
- Encourage communities to construct pedestrian walkways and bicycle paths.
- Create a green city project by planting trees and shrubs.
- Create educational programs in schools and churches to promote environmental issues.
- Become involved in local city programs to control pollution runoff from parking lots, streets, and roofs that pollutes recreational areas, beaches, and rivers.
- Become aware of sewage leaks, oil spills, fertilizers, pesticides, and other chemicals that pollute the waters and earth.
- Volunteer with organizations that clean up and protect wildlife habitats, wetlands, and other endangered areas.
- Become aware of the emissions produced by coal and oil refineries. Start a campaign to reduce the output of these gases.

When Traveling

- Combine your errands into one trip when possible.
- Avoid waiting in long drive-through lines. Park your car and go in.
- When purchasing new vehicles, buy one that is more fuel-efficient.
- Consider using a ride-share program.
- Immediately repair oil leaks in your vehicle.
- Keep your car well tuned and tires inflated. (Your car will use less gasoline.)

- When searching for a new home, look for options such as bike paths, the availability of public transportation, and the possibility of walking to work.
- Take your coffee mug with you when traveling instead of using disposable cups at cafés and restaurants.
- When you drive, avoid quick starts and sudden stops when possible. Drive within the speed limit. These techniques save gas.
- When possible, walk or ride a bicycle.
- Walk, car pool, and use public transportation more often. (By eliminating one 20-mile car trip every week, you'll reduce your annual production of CO_2 greenhouse gases by nearly 1,000 pounds. That is equal to the amount of CO_2 that an acre of rainforest can absorb a year. See Caring for All Creation: http://www.earthministry.org/cfac.htm.)

Though you may think conserving in small ways does not impact the environment, remember that, as the nineteenth-century poem goes, "little drops of water and little grains of sand make a mighty ocean and our glorious land." Change always begins with one willing individual.

A Parable of the Wounded Earth

God created for Job the monarch butterfly that is awakened to life each morning by the warmth of the golden sun. God spoke and there was pristine, sparkling water to quench Job's thirst, and pure, heavenly air to support and nourish Job's body. God created the soil so Job could grow corn and other grains and then dropped spring rains for moisture. God created the pomegranate trees that blossomed with fruit for Job's enjoyment, and he found rest in their cool shade.

However, on a sweltering day, just by chance, Job encountered God as he parked his huge, oil-guzzling Hummer in the parking lot of Mount Moriah Church. For months, Job had been despondent regarding some issues affecting his family. Job seized upon this chance to question God about the future.

"God, what have you done to me and my family? Every day there is a calamity of some sort. The hurricane winds demolished our house; two years of downpours have waterlogged my son's fields. You know how I love to hunt, but I find no deer or quail in the forest. And God, it is so hot that my flowers fade, then sag before dying, and the cornstalks droop and wither and fail to produce corn, and the glaciers are thawing, producing great quantities of water that are flooding most of the coastal areas. God, the kudzu and poison ivy have obscured the hiking trails so we can no longer journey in the few woods we have left. I cannot eat the fish because the waters are contaminated, and the smog is so thick that our asthmatic daughter cannot play outside. God, what have we done to deserve all this? I and my fellow chemists and engineers have worked long, hard hours creating vehicles, factories, concrete, plastics, aluminum, insecticides, and thousands of other chemicals in order to make life easy, enjoyable, and—most of all—profitable. Why should I now suffer? Do you not care about us anymore?"

"My son," God replied, "you have wounded the earth, and your bill has come due. The earth I created was good, but now I hardly recognize my own creation. The assaults made upon the earth, humans, animals, and the ecosystem have caused much of your anguish. You have brought destruction upon yourself. You and only you can rectify that which you have done."

Questions for Reflection

1. What can you do to promote creation care in your home, church, and community?
2. What is the implication in Isaiah 43:16-21 regarding the Lord's proclamation, "I'm going to do a new thing"?
3. How will your environmental footprint become part of your legacy?
4. A coalminer said, "We think we can wound the earth and walk away, but one day the bill will come due." Has that time come, or is it possible to reverse or slow the magnitude of the debt we owe the earth?

WEBSITES PROMOTING A HEALTHIER ENVIRONMENT

sierraclub.org: Oldest and largest environmental organization

savebiogems.org: National Resource Defense Council

earthshare.org: Environmental groups working together under one name

greenhomeguide.com: Natural home materials

energystar.gov: Energy-efficient appliances

mbayaq.org: Seafood watch at the Monterey Bay Aquarium

environmentaldefense.org: Environmental Defense Fund

zerowasteamerica.org: Zero Waste America

travelgreenwisconsin.com: Travel Green in Wisconsin

AmericanProgress.org: Search "global warming"

NationalGeographic.com: Learn about the many facets of our exquisite planet arborday.org: National Arbor Day Foundation, caring for and planting trees

greenhousenet.org: Green House Network

NWF.org/globalwarming: National Wildlife Federation

greenerchoices.org/eco%2Dlabels/: Consumer Reports Greener Choices Guide to Environmental Labels

greenerchoices.org: Consumer Reports endorses products for a better planet

UCSUSA.org: Union of Concerned Scientists

abundantforests.org: Ways to preserve the forest

BOOKS RELATED TO SAVING THE ENVIRONMENT

Harm J. de Blij, Maller and Williams. *Physical Geography: The Global Environment Text Book and Study.*

Tim Flannery. *The Weather Maker: How Man Is Changing the Climate and What It Means for Life on Earth.*

Sheherazade Goldsmith. *A Slice of Organic Life.*

Al Gore. *An Inconvenient Truth: The Planetary Emergency of Global Warming and What We Can Do About It.*

Paul Hawkins. *Blessed Unrest.*

John Kerry and Teresa Heinz Kerry. *This Moment on Earth.*

Daniel A. Kriesburg. *A Sense of Place: Teaching Children about the Environment with Picture Books.*

Paul Simon. *Tapped Out: The Coming World Crisis in Water and What We Can do About It.*

Alan Snitow and Deborah Kaughman. *Thirst.*

VIDEOS AND FILMS ON CREATION CARE

Too Hot Not to Handle—global warming primer; HBO documentary; see http://www.hbo.com/docs/programs/toohot/index.html

An Inconvenient Truth—creation care as a top priority; see http://www.climatecrisis.net/

The Great Warming—encourages environmental stewardship; see http://www.thegreatwarming.com/

A Crude Awakening—see oilcrashmovie.com

King Corn—see kingcorn.net for "a feature documentary from mosaic films incorporated"

See these videos from "The Video Project" at www.videoproject.com:
   Crude Impact—global dependence on fossil fuels
   Environmental Ethics
   Dare to Be Aware Series
   The Man who Planted Trees

See these videos from Bullfrog Films at www.bullfrogfilms.com:

Offluenza—disease of materialism and guide to simple living

The Air We Breathe

America's Lost Landscape: The Tallgrass Prairie—alterations of nature

Anna and the Rain Forest

Baked in Alaska—a look at Arctic National Wildlife Refuge

Printed in the United States
137016LV00003B/9/P